THE ULTIMATE
CIGAR
ENCYCLOPEDIA

THE ULTIMATE
CIGAR
ENCYCLOPEDIA

JULIAN HOLLAND

CONSULTANT: NEIL MILLINGTON

LORENZ BOOKS

First published in 1998 by Lorenz Books

© Anness Publishing Limited 1998

Lorenz Books is an imprint of
Anness Publishing Limited
Hermes House
88–89 Blackfriars Road
London SE1 8HA

This edition published in the USA by Lorenz Books
Anness Publishing Inc., 27 West 20th Street,
New York, NY 10011; (800) 354–9657

This edition is distributed in Canada by Raincoast Books,
8680 Cambie Street, Vancouver, British Columbia, V6P 6M9

ISBN 1 85967 755 X

A CIP catalogue record for this book is available from the British Library

Publisher: Joanna Lorenz
Senior Editor: Lindsay Porter
Editorial Assistant: Kathrin Henkel
UK Research: InfoSearch
US Research: Shannon Ryan, Amy Wilensky
Designer: Nigel Partridge
Photographers: Walt Chrynwski, John Freeman, Dave Jordan

Printed and bound in Hong Kong

1 3 5 7 9 10 8 6 4 2

ACKNOWLEDGEMENTS
The Publishers would like to thank the following for their assistance in producing this book:
Neil Millington of the Havana Club; Alfred Dunhill Ltd; Desmond Sautter of Sautter's of Mayfair;
Tim Cox of J.J. Fox's; Harrods Cigar Room; The London Cigar Importing Co. Ltd; N. R. Silverstone Ltd;
Premium Cigar Ltd; Rothmans International; Swisher International Ltd; Valdrych Cigars.

CONTENTS

INTRODUCTION

Cigars are smoked slowly, for relaxation and enjoyment, or for their flavour and aroma during contemplative moments. Serious cigar smokers have strong preferences for particular brands and loyalty to the shops that supply them. In the early days of World War II, Winston Churchill received a message from the manager of Dunhill, whose shop had just been bombed: "Your cigars are safe, Sir". It is rumoured that prior to the US trade embargo on Cuba, President John F. Kennedy ordered stocks of his favourite Havanas.

Both of these anecdotes reveal the importance of the cigar to the connoisseur: they are appreciated, discussed and studied in much the same way as a vintage wine or fine whisky. A premium cigar has qualities that depend upon

ABOVE AND LEFT: A cigar cutter and amber cheroot holder. Early smoking accessories are now highly collectable.

BELOW: Bundles of premium cigars, grouped according to colour.

ABOVE: Elaborately painted cases for "segars", as they were called in the 19th century.

where the tobacco was grown, how it was fermented, the blend of the leaves and the skill of the makers – each stage in the process is handled by an expert in that particular skill, whether stripping the leaves or rolling the tobacco.

Because of the time and expertise required to create high-quality cigars, they are not cheap – this does not mean that only the very wealthy smoke cigars, however. Recent years have seen a renaissance in cigar smoking and appreciation, with specialist shops, clubs and restaurants devoted to the pleasures of smoking cigars. For the newcomer to this fascinating world, the wide variety of cigar types may seem bewildering. This book aims to demystify the subject and to enhance the enjoyment and pleasure of anyone who appreciates a cigar.

RIGHT: A metal case for carrying a single cigar, from the late 19th century.

TOBACCO

Tobacco, the leaves of the plant *Nicotiana tabacum*, came to Europe from the Yucatan peninsula in Mexico in 1558. It was brought by the early explorers of the New World and was popularized in France by Jean Nicot (1530–1600), the agent of the King of France in Portugal, at that time. His name is commemorated in the tobacco plant itself and the substance, nicotine, that is found in all tobacco. The first tobacco plants were introduced to England by Sir Walter Raleigh in 1570, and legend has it that he was the first Englishman to take up the habit of smoking. The celebrated story relates that his servant found smoke arising from his master's head and threw the contents of a tankard over him to douse the fire.

THE ORIGINS OF TOBACCO SMOKING

It is impossible to say when tobacco was first smoked, but the tobacco plant was certainly grown by the Mayans in Central America and smoked in religious ceremonies. The habit may have spread from there, northwards to Mexico and the Yucatan peninsula. Alternatively, the habit may have spread southwards from Mexico. Certainly tobacco pipes have been found in prehistoric sites in Latin America and, when the Mayan civilization was destroyed following conquest by the Spanish

"A cigar numbs sorrow and fills the solitary hours with a million gracious images."

GEORGE SAND

in the 16th century, the Mayan people dispersed throughout the countries of that continent. They took with them their habit of smoking and also the seeds of the tobacco plants. The Mayan word for smoking is *sikar,* which eventually became *cigarro* in Spanish.

The transatlantic voyage of Christopher Columbus in 1492 was followed by the conquest of Central America by the Spanish conquistadors. Notable among these were Vasco Nunez de Balboa, who discovered the Pacific and founded a colony at Darien; Hernando Cortes who, starting from Cuba, overthrew Montezuma and won Mexico from the Aztecs; and Francisco Pizarro who invaded the Inca kingdom of Peru in 1532. In the course of these conquests the conquistadors found the natives smoking the tobacco leaf, chewing tobacco and even taking it as snuff. The natives of America looked on smoking as a sacred tribal custom, symbolized by the peace pipe. This practice of "drinking smoke" was adopted by sailors and conquistadors, who brought the tobacco back to Europe, where it was regarded as one of the marvels of the New World. Smoking was first looked on as a medicine, then a luxury, and then as a fashion. Set by Sir Walter Raleigh, this fashion spread rapidly throughout Europe, helping to create a demand for the plant that was soon to bring wealth to the early colonists of Central and North America.

BELOW: A 19th-century colour print depicting Christopher Columbus setting foot on Central American soil in 1492.

ABOVE: Mayan stone carving, found in the ruins of Copan, depicting Butz Chan or the "Smoke Serpent".

THE FIRST SMOKERS

From the time of its earliest introduction into Europe, there was a great outcry against the use of tobacco. King James I of England published *A Counterblast to Tobacco,* a bitter pamphlet against its use. He branded it "a custom loathsome to the eye, hateful to the nose, harmful to the brain, dangerous to the lungs, and in the black, stinking fume thereof, nearest resembling the horrible Stygian smoke of the pit that is bottomless", and a great deal more in that vein. His opposition proved fruitless and the popularity of the "queen herb of the rude barbarian", as it was described in China, has steadily increased ever since the 16th century.

LEFT: *A leather tobacco pouch, silver-mounted clay pipes and a bone tobacco stopper, believed to have once been the property of Sir Walter Raleigh.*

ABOVE: *A portrait of Sir Walter Raleigh, the man credited with introducing tobacco to England, following his expeditions to North America .*

BELOW: *Seville, depicted in this early 20th-century stage set for Bizet's* Carmen, *was the centre of cigar manufacturing in Spain from the early 18th century.*

The first tobacco smokers, including Sir Walter Raleigh, smoked pipes, and pipe smoking continued in Europe for over a century. At that time it was the only way that tobacco was smoked. But cigars, which originated in the Spanish colony of Cuba, spread from there to Spain at the beginning of the 18th century. Cigars, made from imported Cuban tobacco, were manufactured in Seville, in Spain, from 1717 onwards, and by the end of the 18th century cigars were also being made in France and Germany.

The first cigars were introduced into America by Israel Putnam, an officer who had served in Cuba in the British Army. Putnam later became a general in the American army and fought against the British in the American War of Independence (1775–83).

FROM PIPES TO CIGARS

The use of cigars spread from Spain and Portugal to the rest of Europe, fuelled in part by the Peninsular War (1806–12) against Napoleon. Both the British and French soldiers acquired the habit of smoking cigars from the Spanish, and as cigar smoking and snuff taking became fashionable, so pipe smoking declined. The manufacture of cigars, or "segars" as they were called originally, started in Britain in 1820 and was regulated by an Act of Parliament in 1821. Around the same time, cigarettes became a cheaper alternative to cigars, but it was not until cigarette-making machines were introduced in the 19th century that smoking cigarettes became popular.

BELOW: By the late 19th century, cigar smoking had become commonplace among the leisured classes.

ABOVE: Captains Middleton, Price and Koet, seen here at an inn in London, were among the first to smoke tobacco in England.

THE TOBACCO PLANT

Tobacco can be grown in most climates, ranging from temperate to tropical, but various regions are specially adapted for growing certain types. Some regions produce tobacco for cigars, others produce a "manufactured" kind that is used to make snuff, cigarettes and pipe and chewing tobacco.

Although nearly all the varieties of tobacco come from two or three native American species, tobacco plants differ greatly in size, thickness and the colour of their leaf, according to varying conditions of soil and climate. The plant is coarse and rank, and has large, drooping leaves springing from a thick central stem. The leaves are gummy and clammy to the touch, because of the sticky secretion of the short hairs that cover the green parts of the plant. The flowers – large, sweet-scented blossoms – appear in a cluster at the top of the stalk, and range in colour from deep pink to nearly white. The flower buds are usually cut off before they open, so that all the strength of the plant goes into the nourishment of the leaves. The seeds of the plant are black and so small that 60,000 of them make only one tablespoonful – enough to sow 83.5 sq m (100 sq yd) of seed-bed.

ABOVE: Rows of week-old wrapper plants at a plantation in San Luis, Cuba. The region is famed for producing some of the finest cigar wrappers in the world.

LEFT : Rows of Burley tobacco plants.

RIGHT: In cigar-leaf production, the flowers of the tobacco plant are usually cut off before they open, thus ensuring that all the nourishment goes into producing healthy leaves.

CURING TOBACCO

The tobacco leaf is green when harvested and does not have the characteristic colour and flavour until it is cured, fermented and aged. There are three methods of curing (drying) tobacco. Sometimes the leaves are sun-dried, which results in a very sweet chewing tobacco. The second method is air-drying. This takes place in specially constructed barns, where the ventilation can be carefully regulated. Tobacco used in cigar manufacture is usually air-dried. The third method is by artificial heat – over open fires or in flues. Open fires give the tobacco a characteristic smoky odour. In flue-curing, the fires are outside the tobacco sheds and the heat is brought in through flues or iron pipes. This results in the bright yellow leaf, largely used for the manufacture of light pipe and cigarette tobacco.

After drying, the tobacco leaves are brittle and cannot be handled without crushing them to a powder, so they are left hanging until rainy weather arrives, when the leaves absorb moisture and become as soft and pliable as a kid glove. The leaves are then stripped

"As you approach 30, you have a 30 ring gauge; as you approach 50, you have a 50 ring gauge."

CUBAN SAYING

from the stem, sorted, made up into small bunches, and fermented by piling them in stacks 1.5–1.8m (5–6ft) high. When fermentation is completed, the tobacco is graded and packed. It is then sometimes aged in a warehouse for as long as four or five years, to make it more mellow in flavour.

Often the tobacco is flavoured in manufacture, and sugar, liquorice, spices or alcohol are added to give it an artificial aroma. It goes without saying that taxes on tobacco increase its price considerably, and are a most important source of revenue in many countries throughout the world.

NICOTINE

Tobacco owes its sedative and habit-forming powers to the drug nicotine, a deadly and dangerous poison which is present in all tobacco. Smoking in general has become much less popular in the last 20 years, especially in North America, since the publication of the American Surgeon General's report on its effects on health in the early 1960s. Nevertheless, the Surgeon

ABOVE: *Jean Nicot did much to spread the popularity of smoking in Europe.*

LEFT: *A curing barn in Cuba.*

ABOVE AND RIGHT: Drying tobacco leaves in Pennsylvania, USA (above) and Cuban cigar wrapper leaves after 50 days of air-curing (right). After harvesting, the leaves are sewn in pairs and hung over poles. Freshly harvested leaves are stacked near the floor, and as they dry out are moved higher up in the curing barn.

General's *Report on Smoking* stated that cigar smokers, who smoked fewer than five cigars each day, had the same mortality rate as non-smokers. Since the early 1990s there has been a major revival in the popularity of handmade cigars. They have become fashionable once again, due to the enthusiasm shown for them by film stars such as Jack Nicholson, Arnold Schwarzenegger and Sharon Stone.

RIGHT: An engraving from the late 16th-century book, Traicte du Tabac ou Nicotiane, *depicting the Native American style of processing tobacco.*

THE HISTORY OF CIGARS

As we have seen, cigars similar in form to those we know today were first manufactured in Spain in the early 18th century. By the end of the century, cigar manufacture had spread to France and Germany, and at the same time the Dutch started making cigars, using tobacco imported from their Far Eastern colonies of Java and Sumatra. The manufacture of "segars", as they were then called, commenced in Britain in 1820 and was swiftly followed by an Act of Parliament that regulated production. An import tax on luxury goods meant that premium foreign cigars were a luxury item, and this restricted their popularity for a time. In the United States, domestic production and a levy on tax by 1870 meant cigars were more accessible.

CIGAR SMOKING IN EUROPE

As people abandoned snuff, cigars became more popular and demand sprang up for the higher-quality Cuban cigars. King Ferdinand VII of Spain decreed in 1821 that the production of cigars in Cuba, then a Spanish state monopoly, should be increased. Soon cigar smoking became so widespread that special smoking rooms were introduced in hotels and gentlemen's clubs. Later in the century, special smoking cars were included on many European luxury trains. The after-dinner cigar, accompanied by a glass of brandy or port, became a tradition. The acceptance of this ritual

LEFT: *The after-dinner cigar became a ritual among the highest circles of European society, as depicted in this late 19th-century poster.*

BELOW: *Enjoying an after-dinner cigar at the restaurant of Les Ambassadeurs Hotel in Paris. From a late 19th-century painting by the French artist, Jean Beraud.*

BELOW: *King Ferdinand VII of Spain was responsible for the increase in production of Cuban cigars and their importation into Europe in the early 19th century.*

amongst the highest circles of European society was helped by the fact that the Prince of Wales, later King Edward VII, was an avid cigar smoker. This was despite his mother, Queen Victoria, having a particular dislike of the habit! At the end of the 19th century, Britain formed the largest market for Cuban cigars in the world. However, it was not until the intro-duction of machine-made cigars in the 20th century that the smoking of cigars became more available to the general population.

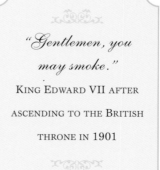

"Gentlemen, you may smoke."
KING EDWARD VII AFTER ASCENDING TO THE BRITISH THRONE IN 1901

CIGAR SMOKING IN NORTH AMERICA

In North America the first tobacco plantations were set up by European settlers in the early 17th century and, until the arrival of Cuban cigars in the mid-18th century, tobacco was only smoked in pipes. When Israel Putnam returned home to Connecticut from Cuba, he brought with him a selection of Havana cigars and large amounts of Cuban tobacco seed. Connecticut was one of the tobacco-growing areas of North America, where plantations had been established by the early settlers, and the new, improved seed provided a fillip to the local tobacco industry. By the early 19th century cigar factories had opened near Hartford, Connecticut, using tobacco leaves grown from the imported Cuban seed. This was the start of a flourishing industry, and today Connecticut produces many of the best wrapper leaves found outside Cuba. At the same time, the tobacco merchants started to import Cuban cigars into North America, and this trade gradually increased as cigar smoking became more popular in the USA.

John Quincy Adams, the sixth president (1825–9), was a confirmed cigar smoker, as was Ulysses S. Grant. Grant became president in 1869 and was elected for a second term in 1873. By that time cigar smoking had become widespread, and following tax

ABOVE: Cigars have been a symbol of power since the early 20th century, as illustrated by this turn-of-the-century cartoon depicting US President McKinley.

reductions in the 1870s cigars became more readily available and domestic production increased. By the end of the 19th century, as in Europe, the cigar had become a major status symbol. To combat the popularity of cigarettes, cigar manufacturers introduced machine production in the 1920s, and this reduced considerably the amount of handmade cigars produced in the USA.

LEFT: An 18th-century print depicting the tobacco trade in the (then) British colony of Virginia.

RIGHT: An American cigar box design with a cowboy theme, circa 1900.

CIGAR MANUFACTURE IN CUBA

Commercial tobacco-growing in Cuba, the home of Havana cigars, began as early as the 16th century. As exports of the tobacco leaf to Spain increased and tobacco farming became more profitable, the tobacco-growing peasants with small plantations were at constant loggerheads with the big landowners. In the 1850s, when there was considerable free trade in tobacco, there were nearly 10,000 tobacco plantations and over 1,000 cigar factories on the island. But, as cigar production became a major industry, the larger landowners forced the peasants out, and by the turn of the century the number of factories had been reduced to 120. Eventually the peasants either became tenant farmers or had to move

> *"What this country needs is a good five-cent cigar."*
>
> THOMAS MARSHALL, US VICE PRESIDENT, 1919

to other areas of the island. The majority of the cigars were exported to the USA and Spain.

In the last half of the 19th century, cigar production in Cuba became more sophisticated, and brand and size differentiation began. Many of the distinctive cigar bands and boxes that we know today were introduced at this time. However, tariff barriers were brought in by the USA in 1857, reducing the market for Cuban cigars. This led to Cuban cigars finding new markets in Europe.

BELOW: A tobacco plantation in the Pinar del Rio province of western Cuba – one of the finest cigar-growing regions in the world.

ABOVE: A Cuban tobacco farmer, or veguero, *is supplied with tobacco seed by the Tobacco Research Station to ensure a good crop.*

ABOVE: An illustration depicting Havana cigar-makers demonstrating against the mechanization of their trade.

As the 19th century drew to a close, Cuba started to rebel against the oppressive regime of Spain as a colonial power, and the struggle for independence began. Many cigar-makers left Cuba and settled in Florida and Jamaica. After the short Spanish-American War of 1898, the island came under the influence of the USA and cigar exports again flourished. In 1933 Fulgencio Batista y Zalvidar, then a sergeant-major in the army, was instrumental in the coup against President Machado. Batista was president from 1940–4 and was re-elected in 1954. He ruled as dictator until his overthrow by Fidel Castro in 1959.

THE CUBAN CIGAR INDUSTRY UNDER CASTRO

The Cuban revolution of 1959 changed the nation's cigar manufacture. The cigar industry, much of it American-owned, was nationalized and became a state monopoly operating under the name of Cubatabaco. Following the introduction in the USA of inexpensive machine-made cigars, using homogenized tobacco leaf, the Cubans hit back by also producing machine-made cigars.

After the revolution, many important cigar-factory owners fled abroad and set up new operations in the Dominican Republic, Florida, Honduras, the Canary Islands and Mexico. The USA broke off relations with Cuba in 1961 and imposed a trade embargo the following year. This meant that Havana cigars could not be legally imported into the USA, except for small quantities intended for personal use.

The remains of the Cuban cigar industry have had mixed fortunes in recent years, with several bad harvests and falling production in the 1990s. The collapse of Communism meant a severe shortage of Eastern bloc materials. In an attempt to inject new life into the Cuban cigar industry, a new company, Habanos SA, took over the marketing of Havana cigars in the mid-1990s.

LEFT: Fidel Castro nationalized his country's cigar industry soon after ousting the Batista regime in 1959.

CIGAR MANUFACTURE OUTSIDE CUBA

The Cuban émigrés, with the help of American backers, such as the Consolidated Cigar Corporation and General Cigar Company, have now built up a major cigar industry in other Central American locations. Half of all handmade cigars imported into the United States are now manufactured in the Dominican Republic, a country with a similar climate and tobacco-growing conditions to Cuba. Other countries which have benefited from the migration of Cuban cigar manufacturers are Honduras, Nicaragua, Jamaica and Mexico, as well as Florida in the United States.

"Do not ask me to describe the charms of reverie, or the contemplative ecstasy into which the smoke of our cigar plunges us."

JULES SANDEAU

Apart from areas of Central America, many other countries around the world either produce top-quality cigars or the raw materials for their production. These include the United States, Ecuador, Brazil, the Canary Islands and Cameroon. The Indonesian islands of Java and Sumatra also supply raw materials to the manufacturers of machine-made cigars in Germany, Switzerland and Holland, while the Philippines supplies Spain.

BELOW: *Harvesting tobacco in the United States. Bundles of leaves, seen on the trailer, are then taken for curing.*

ABOVE: *A tobacco farmer and his newly planted crop in the Vallarta region of Mexico.*

LEFT: *Sorting bundles of handmade cigars in a Nicaraguan factory.*

THE CIGAR TODAY

Even at the end of the 20th century, when the mass-production of consumer goods is the norm, the manufacture of handmade cigars is still a very labour-intensive operation. Good cigars have distinctive qualities that depend on where the tobacco was grown, the fermentation process and the skill of the maker. In fact, good quality hand-made cigars can be likened to fine wine. Because of the time required for cigar tobacco leaves to mature and the number of hours involved in their manufacture, they are neither cheap to buy, nor in plentiful supply. Choose them with thought and take care when storing them. When smoked, they should be savoured slowly for their flavour and aroma.

A HAVANA CIGAR

hemically, tobacco is split into two groups. The main group is saccariferous tobacco, so-called because it contains sugar, which causes an acid reaction to the smoke. This tobacco is used for cigarette manufacture. Cigar tobacco is alcalic or basic tobacco and produces a neutral smoke. It is air-cured, with fermentation taking place naturally.

Tobacco will grow in many climates, ranging from temperate to tropical, but various regions are especially suitable for growing certain types. As with the finest wine, producing the best cigars depends on the type and quality of the tobacco leaves, the soil in which they are grown, the aspect of the plantation, the weather conditions in each year, the way the leaves are allowed to ferment and, finally, the skill and expertise of the maker.

The best cigars have three parts that use different parts of the tobacco plant: these are the filler, the binder and the wrapper. In Cuba, five types of tobacco leaf go to make a Havana cigar, commonly acknowledged as the best in the world. The *tripa*, or filler, contains a blend of three leaves: *ligero*, *seco* and *volado*. These are secured by the *capote* (binder) while the *capa*, or wrapper, dresses the cigar and dictates its appearance.

There are two seed varieties, each of which produces distinctly different tobacco plants, known as *corojo* and *criollo*. The *corojo* plant, named after El Corojo Vega, the famous plantation where its seed was developed, is designed to produce the finest *capa* or wrapper leaves,

"If I cannot smoke in heaven, then I shall not go."
MARK TWAIN

and is the most expensive tobacco used in Havana cigars. It is grown under gauze sheets held up on tall poles.

The leaves on a tobacco plant are grouped into seven levels on the stem. From top to bottom these are: *coronas, semi-coronas, centro gordo, centro fino, centro ligero, uno y medio* and *libra de pie*. Wrapper leaves are also classified by colour, in seven shades, although in the past manufacturers have used many terms for wrapper leaves, some of which are still in use. These are: *double claro* (light green), *claro* (light tan), *colorado claro* (light to mid-brown), *colorado* and *colorado maduro* (reddish-brown to medium brown), *maduro* (dark brown) and *oscuro* (dark or black).

LEFT: *A Montecristo Especial No 2, a fine Cuban brand.*

RIGHT: *The three main components of a handmade cigar: wrapper, filler and binder.*

wrapper   binder   filler

GROWING TOBACCO

The *criollo* plant, the perfected strain of the only true Cuban tobacco seed, produces four of the five tobacco leaves which are blended to create the many different flavours found in Havana cigars. *Criollo* plants, with their six or seven pairs of leaves, are exposed fully to the sun to help produce the wider variety and greater intensity of flavours for the different blends used in Havana cigars. The best *criollo* plantations are known as *Vegas Finas*. The roots of tobacco plants are delicate and require the loosest possible soil if they are to grow successfully. During the summer months the plantation fields are ploughed several times. Tobacco plantations have to be situated on flat, well-drained land, and animal power is used for this operation, to ensure that the soil is not compacted in cultivation. Tobacco seeds are then provided, free of charge, by Tobacco Research Stations to the tobacco farmers.

Growing tobacco is an extremely labour-intensive operation, with each plant being tended over 150 times during the 120-day cycle. Following the planting of the seeds – in Cuba this is usually in the middle of September – they are covered with straw or cloth to protect them from the heat of the sun. As the seeds germinate they are sprayed with pesticides and after

ABOVE: A field of tobacco plants ready for harvesting in South Carolina, USA.

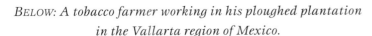

BELOW: A tobacco farmer working in his ploughed plantation in the Vallarta region of Mexico.

about 45 days, when their height is 15–20cm (6–8in), towards the end of the rainy season, the seedlings are transplanted into the tobacco fields. The leaves are watered both by rain and the morning dew, and irrigated from below. As the leaves develop, the buds appear and are removed by hand to encourage plant growth.

The best *corojo* wrapper leaves come from tobacco plants grown under a canopy of muslin sheets supported by wooden poles. This keeps the leaves smooth and prevents them from becoming too oily. Shade-grown tobacco, such as the finest Connecticut Shade, is grown under tall, tent-like structures of cheesecloth or mesh.

HARVESTING TOBACCO

At harvest time, around 80 days from the planting of the seed, the leaves are picked by hand, using a single movement. This is a very labour-intensive operation. The leaves are cut in six stages starting at the bottom of the plant, each stage taking about a week. Broadly speaking, the highest-quality leaves are found in the middle, or *centro,* part of the plant, and are used as wrappers. If all the leaves are in good condition, each plant can wrap over 30 cigars. The upper, or *corona,* leaves are used as fillers, and the lower leaves as binders. The blend of these leaves, encased in an appropriate wrapper, is what makes a given cigar mild or medium bodied.

When the leaves have been picked they are threaded together in bundles and hung up in batches of 50. The piles are strung up on horizontal poles and cured in large barns, called *casas del tabaco* in Cuba, near the tobacco plantations. The curing barns are built facing

LEFT: Plantation workers harvesting tobacco at Montpelier, Jamaica, circa 1900.

west so that one side of the building is heated in the morning and the other in the afternoon. The wooden poles, on which up to 100 leaves of tobacco are suspended, are held horizontally high in the barn. This allows air to circulate. This drying process takes between 45 and 60 days to complete, depending on the weather. The heat and humidity in the barns are closely monitored and the green leaves turn to their characteristic brown as the chlorophyll in the leaves turns to carotene.

Once the leaves are dried the poles are lowered and the leaves are sorted by size, texture and colour. They are then made up into bundles of 20 leaves called "hands" and taken to the fermentation house. However, a small amount of moisture is left in the leaves so that fermentation can take place.

BELOW LEFT AND RIGHT: Growing and harvesting tobacco in Cuba, in the Valle de Vinales region (left) and in the Pinar del Rio region (right).

FERMENTATION

Fermentation of cigar tobacco takes place in two stages. In the first stage the leaves are heaped up in piles about 1m (3ft) high, under jute burlap coverings, and left to ferment for up to three months. During this period it is essential that the temperature of each pile does not exceed 33°C (92°F) and the colour of the leaves assumes a uniform shade. The piles of leaves are then broken up, moistened with water to facilitate handling, and graded in the sorting house for use as wrapper, filler or binder. They are also sorted according to size, colour and quality.

The flattened leaves are then returned to the fermenting house, stacked in large piles up to 2m (6ft) high on boards, tightly packed and stored in darkened rooms. In this state they are known as *burros*. The leaves now undergo a secondary fermentation for up to two months, where the temperature must not exceed 43°C (110°F). The temperature is monitored by long, sword-like thermometers which are thrust into the heart of each *burro*. If the temperature rises too high, the piles are broken down and the leaves are

ABOVE: Harvested tobacco leaves being hung on poles, ready for curing, in Ecuador.

restacked with the top becoming the bottom. Rebuilding the *burros* may take place as often as ten times during the secondary fermentation. This second process allows the impurities of moisture, sap, nicotine and ammonia to be released from the leaves, and because of this, the level of tar, acidity and nicotine is much lower in cigars than in cigarette tobacco. The exact amount of time spent depends on the type of leaf being produced, and it may take as long as six months for the darkest *maduro* wrapper leaves to achieve their rich shades of black and brown.

MATURING THE TOBACCO

Finally, the leaves are sorted again, wrapped in palm leaves in square bales called *tercios*, and sent to the warehouses, where they are allowed to mature slowly – in some cases for as long as two years, depending on which brand they are destined for. The length of time it takes to mature high-quality tobacco leaf is not the only reason for the high prices of premium cigars. The care taken over maturing leaves properly is one of the key factors determining the quality of a cigar.

BELOW: Nicaraguan tobacco pickers carefully sorting leaves, following harvesting at a plantation near Jalapa.

THE STRUCTURE OF A HANDMADE CIGAR

There are three parts that go to make up a handmade cigar: the wrapper leaf, the binder leaf and the blend of filler leaves. Each part has a role in the manufacture and smoking of a cigar.

THE WRAPPER

The outside wrapper is the most important – and the most expensive – element of a cigar and dictates the cigar's appearance. In some cigars it can account for 60% of the taste. The tobacco leaf used must be smooth, not too oily, and have a subtle bouquet. It also has to be pliable, with no protruding veins, for ease of handling by the roller. Wrapper leaves are normally matured for up to 18 months. Handmade Cuban cigars use home-grown wrapper leaves. However, in other countries wrapper leaves can be supplied from many parts of the world, including Mexico, Nicaragua, Cameroon, Indonesia, Honduras, Ecuador and from the USA, the much-prized Connecticut Shade. The rolling of wrapper leaves is the most demanding part of the

"If you forget a line all you have to do is stick the cigar in your mouth and puff on it until you can think of what you've forgotten."

GROUCHO MARX

making of a handmade cigar. The roller, known as a *torcedor* in Cuba, is the most skilled member of the team, and it can take up to 20 years to become a master roller.

CUBAN WRAPPERS

The delicate nature of Cuban wrappers demands extra care to restore the supple silkiness of the tobacco leaves, before they are given a final classification and sorting. This is achieved by a special *moja,* or moistening, performed during the cool, early morning hours. Water is shaken out and the leaves hung up overnight

BELOW: Wrapper leaves after being hung for 50 days of air curing.

BELOW: The structure of a handmade cigar: with wrapper (top); without wrapper (middle); the filler with binder removed (bottom).

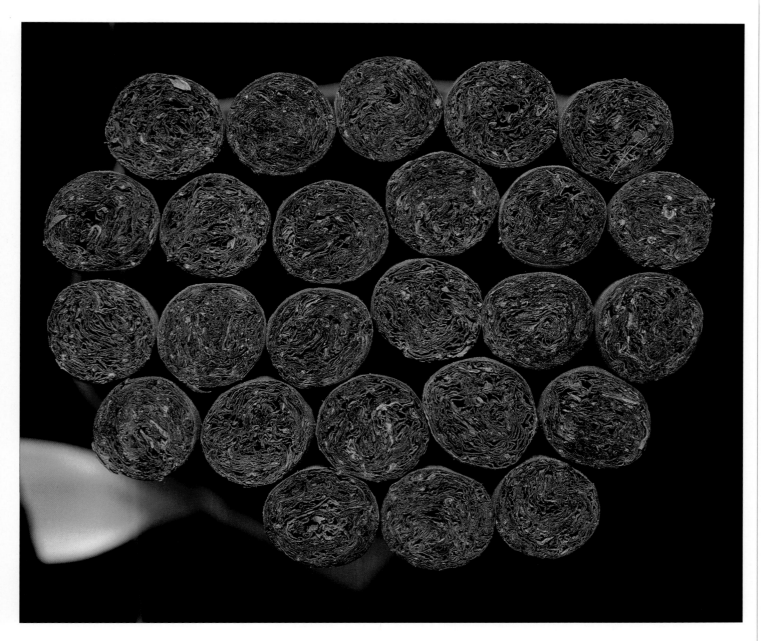

ABOVE: End-on view of H. Upmann Connoisseur cigars, demonstrating their structure.

to ensure that the moisture is even. The *despalilladoras,* or strippers, remove the stems whole, bisecting the leaves. Finally the *rezagadoras* (graders) sort them into piles according to size, colour and texture before they move on to the next stage. The *rezagadoras* work from piles of leaves on their laps, perhaps giving credence to the myth that Havanas are rolled on maidens' thighs.

THE BINDER

The binder leaf holds the filler leaves together inside the cigar. It is usually a coarse leaf, harvested from the upper part of the tobacco plant, and is selected for its physical strength. The binder also affects the burn-rate, aroma and taste of a cigar, and it is vital that its flavour must not be incompatible with that of the wrapper and filler. Processed binders are used for many machine-made cigars.

LEFT: Early cigar box label celebrating the tobacco leaf.

ABOVE: A rezagadora sorting wrapper leaves by size, colour and texture in the Partagas factory, Havana.

BELOW: Bunches of filler and binder tobacco are placed into wooden moulds and then pressed.

ABOVE: Cured wrapper leaves, suspended on wooden poles, are closely monitored before fermentation.

THE FILLER

The tobacco leaves used to make up the filler are folded together lengthways in a concertina (accordion) fashion. This forms a series of tiny passages in the cigar, through which air and smoke can be drawn. This folding process is better achieved by hand and is the reason that handmade cigars are better to smoke than machine-made products. A lengthways cross-section of a cigar looks like a cross-section of the pages of a book.

There are three different types of leaf used for the filler. Dark-coloured, strongly flavoured, slow-burning *ligero* leaves, harvested from the top of the tobacco plant, are placed in the middle of a filler. These leaves are matured for two years before being used in cigar manufacture. Lighter-coloured, milder *seco* leaves from the middle of the plant surround the *ligero* leaves. These leaves are matured for a period of up to 18 months. *Volado* leaves, with little flavour, from the lower part of the plant, are placed on the outside of the filler section because of their excellent burning qualities. They are matured for a period of up to nine months. The relative proportions of these three types of filler leaf are the key to the flavour of a handmade cigar. The art of the cigar manufacturer is to keep these proportions consistent within any given brand.

HOW CIGARS ARE MADE

There are seven production stages in the making of a handmade cigar and the process has remained virtually unchanged for the last 150 years. Depending on the size and blend, two to four filler leaves are folded up like a fan and then rolled into the two halves of the binder leaves. It is essential that the filler leaves are evenly distributed as this is what makes a cigar draw evenly. If the

"A Hoyo de Monterrey double corona is my favourite Cuban since Desi Arnaz."

BILL COSBY

cigar is filled too loosely, it will burn too fast and become acrid. If the filler leaves are too tight, the cigar will not draw properly. The filler leaves are important as they are the key to a consistent flavour and draw for each brand. When the blend of filler leaves has been rolled into the binder, the cigar is known as a "bunch".

This is then placed into a wooden mould, about ten in a row, the right

BELOW: *A typical Havana cigar maker's workbench with the flat-bladed* chaveta *in the foreground.*

BELOW: *The construction of a handmade Havana cigar involves seven production stages.*

ABOVE AND BELOW: Two stages in the rolling of a handmade cigar. After the bunches are pressed and trimmed, they are arranged across the wrapper leaf and rolled. When rolling is complete, the end is stuck down with flavourless vegetable gum (glue).

size for the cigar being made. The surplus filler is trimmed from the ends of the bunch and a stack of moulds are placed in a bunch press. The pressing time usually takes about three-quarters of an hour, and during this period the cigars are pressed gently and turned regularly, to ensure a perfect cylindrical shape. After the bunches are pressed and trimmed, they are ready for the next stage of production, when the wrapper leaves are rolled round them.

THE WRAPPER LEAF

An oval steel cutter, known as a *chaveta,* is used to trim the selected wrapper leaf to the right size. The leaf top is placed at the foot of the cigar and the leaf base is at the head. This ensures a milder flavour is experienced during the smoker's first few puffs. The moulded

bunch is then placed at an angle across the trimmed wrapper and rolled from one end. The wrapper is gently stretched and wound around the bunch, with each turn slightly overlapping. After the rolling is completed, the loose end is stuck down using a flavourless vegetable gum (glue). The cigar is then rolled under the flat blade of the *chaveta* to ensure that it has been evenly made. A circular piece of left-over wrapper leaf is cut out to form a cap, which is glued in position using the flavourless vegetable gum. In the manufacture of the most expensive cigars the closed end is actually sealed by twisting the end of the wrapper. In the final process, the open end of the cigar is cut with a sharp guillotine to the correct length. After rolling, cigars are tied up with coloured ribbon in bundles of 50, and are treated for pests in a vacuum fumigator. They are then stored in cedar-lined rooms, known as *escaparates,* to allow them to shed any excess moisture.

QUALITY CONTROL

At the same time as the cigars are being treated for pests, a small percentage of each batch, from each roller, undergoes a rigorous quality-control test. Usually about 10% of the output is tested, with the checklist including length, firmness, weight and smoothness of the wrapper. Afterwards cigars from

BELOW: The weight and size of Havana cigars are carefully checked during rigorous quality-control testing.

each batch are blind-tasted by professional smokers, known in Cuba as *catadores*. This tasting is only carried out during the morning and only about 2.5cm (1in) of each cigar is smoked. The *catadores* must assess how evenly the cigar burns, the draw, flavour and aroma. This quality-control stage is vital, as a premium product is expected by the consumer.

COLOUR GRADING

After three weeks the bundles of cigars are removed from the *escaparates* and sent in batches of 1,000, from each size and brand, to be graded according to colour. This is quite a task, as there are over 60 possible

ABOVE: Applying the distinctive Cohiba band to a much sought-after Lancero cigar – one of Castro's favourites.

shades to select from. Cigars of the same colour are then packed in wooden boxes, ranging from the darkest on the left to the lightest on the right. Finally, the colour-graded cigars are sent to the packing department of the factory, where their distinctive bands are applied, and they are finally sealed into traditional cedarwood boxes. Each box is labelled and sealed as proof of its authenticity. The use of cedarwood boxes goes back to 1830, when the bank of H. Upmann decided to send cigars to its senior staff in London. The choice of cedarwood was deliberate, as it helps to keep cigars moist and allows for further maturing.

BELOW: Partagas Lusitanias are carefully packed into their traditional cedarwood box.

BELOW: Havana cigar rollers, or torcedores, *are kept contented in their work by the reading of edifying novels or government newspapers – a practice which has existed in Cuban cigar factories for over 130 years. The reader is chosen by his workmates.*

MACHINE-MADE CIGARS

Machine-made cigars are cheaper to make than hand-made ones, because the manufacturing process is quicker and less labour-intensive. They are usually made with homogenized tobacco leaf that is inferior and cheaper, and there is less wastage. Homogenized tobacco leaf is a material consisting of tobacco stems and fibres that is mixed with a substance not unlike cellulose. This material is produced in sheets and is used for both binder and wrapper.

The better quality machine-made cigars, such as the popular European brands of Villiger and Schimmelpenninck, are made of 100% tobacco. Most machine-made cigars are not made with fillers that extend the whole length of the cigar. The binders, too, are quite often made of cut-up tobacco that is then put back together again. The wrappers used are usually of a lesser quality than premium. This means machine-made cigars become hotter and burn more quickly than handmade ones. There are some exceptions, but even the finest machine-made cigar is not as good as a handmade example.

At the cheaper end of the machine-made market, cigar production is not dissimilar to that of cigarettes. The filler consists of small, cut pieces of blended tobacco. These pieces are machine-bunched, processed into rods and then covered by a continuous sheet of binder. After being sealed at the correct length, the wrapper is added, followed by the final trimming.

Better quality machine-made cigars have slightly more personal attention paid to their making. The filler, consisting of scraps of

> *"The true smoker abstains from imitating Vesuvius."*
> AUGUSTE BARTHELMY

tobacco leaf, is fed into a hopper, while the cut binder leaves are fed into a rolling machine. This machine wraps the binder around the filler, and out pops a cigar, which is then trimmed to the required length.

It is not difficult to spot the difference between most machine-made and handmade cigars: machine-made wrappers are less smooth and more veiny, and the caps are more pointed. However, many mass-market machine-made cigars do not even have a cap. Some Cuban cigars are sold as "hand-rolled" or "hand-finished" – this only means that the wrapper is applied by hand and the rest of the cigar is made by machine.

However, it should be noted that machine-bunched and hand-rolled cigars, in addition to being cheaper than completely handmade ones, are also consistent in quality and usually quite mild in flavour. Machine-made dry cigars, such as the European brands of Villiger and Schimmelpenninck, are extremely popular throughout the world. Some of these small, dry cigars provide an excellent smoke at a very reasonable cost.

MANUFACTURING THE BUNCHES

If the cigars are made with a binder of leaf tobacco, the bunches are made on machines individually. If homogenized tobacco is used, bunch-making is done in rods, similar to a cigarette machine. The rods are cut into equal lengths and small pieces of tobacco are drilled out of the ends, which are then pressed to form the mouthpiece.

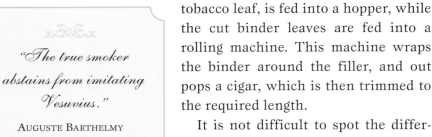

LEFT: *The Dutch brand Schimmelpenninck includes a range of top-quality machine-made cigars.*

ABOVE: Machine-made cigars being checked for quality at the Villazon factory, Florida, USA.

RIGHT: Villiger is a Swiss company producing a well-regarded range of machine-made cigars.

WRAPPING THE BUNCHES, THE CONVENTIONAL METHOD

When the bunches are ready, they are over-rolled with a wrapper leaf on wrapper machines. The operator places the wrapper leaf on a die-cutter which cuts out the required shape. Much practice is required to obtain the maximum number of wrappers out of one leaf. The cut wrapper leaf is then fed into an over-rolling device that wraps the leaf around the bunch.

WRAPPING THE BUNCHES, THE AUTOMATIC METHOD

The preparation of the moistened wrapper leaf and the manufacturing of the bobbin itself, with die-cuts, is carried out in low-wage countries. The graded and humidified wrapper leaf is supplied to an employee, who mechanically cuts the required dies out of the

leaf. Each die is then placed on a bobbin which can carry up to 7,000 dies. This applies both to wrapper and binders and the same type of bobbin is used on both wrapping and bunch-making machines.

A special device, the BUD (Bobbin Unwinding Device), has been developed to unwind and unload bobbins at the bunch and wrapper machines. This device can also detect a damaged binder or wrapper, thus preventing the production of low-quality bunches and cigars.

PACKING

After manufacturing, cigars may or may not be powdered, pressed, cut to size and finally packed. Tobacco is a natural product and every leaf is different. At one time, cigar smokers required uniform colours. This led to the notorious colour sorting, when production runs were often sorted out of 100 different colours. However, due to high wage costs in Europe, this sorting process has now become too expensive. Today, modern machines spread tobacco powder over the cigars to give them a uniform colour and improved burning qualities.

In some European countries, particularly Holland, Belgium and France, there is a growing trend towards so-called natural, unpowdered cigars. This type of cigar is not sorted according to colour and therefore has an untidy or "wild" appearance. These wild cigars have now become fashionable among the younger cigar smoker.

FINISHING

After packing, the cigars are stored in a drying room, where they will be dehydrated in a natural drying process. The moisture, which is essential during production to prevent the leaves from cracking, must now evaporate to restore the original flavour of the tobacco. When the cigars have returned to approximately 12% relative humidity, the packing process will be completed and the cigars are ready.

CIGAR SHAPES

The fatter the cigar, the slower and more smoothly it smokes. Not only can the filler leaves be blended more subtly in bigger cigars, but they are also made by the most skilled rollers in the factory, and although this guarantees their quality, it also inevitably adds to their price.

There are more than 60 different combinations of cigar sizes. The sizes sometimes have the same names, but the exact dimensions often vary from brand to brand and the choice is bewildering. The thickness of a cigar is referred to in terms of its ring gauge, which is expressed in measures of $\frac{1}{64}$in(0.4mm). So if a cigar has a ring gauge of 32, it is $\frac{32}{64}$in or $\frac{1}{2}$in (12.5mm) thick; and if a cigar has a ring gauge of 64, it is 1in (25.4mm) thick.

The beginner can be confused, because different manufacturers use the same name to denote different types of cigar. There are two basic shapes: straight cigars (*parejos*) and irregularly shaped cigars (*figuardos*). Straight-sided cigars include the following shapes:

CORONAS These set the standard against which most cigars are measured. They measure 14–15.2cm (5½in–6in) x 42 or 44 ring gauge. They all have an open foot (the end of the cigar you light), and a closed head (the end you smoke). Variations within this group include: double corona (17.8cm/ 7in x 49 ring gauge), which is slightly

"He who doth not smoke hath either known no great griefs, or refuseth himself the softest consolation, next to that which comes from heaven."

E. G. BULWER-LYTTON

longer and fatter than a corona; Churchill (17.8cm/7in x 47 ring gauge), a strong, full-bodied cigar named after Winston Churchill; and robusto (12.7cm/5in x 50 ring gauge), a short fat cigar.

PANETELAS These are generally longer than coronas and a good deal thinner. The standard measurement is 17.8cm (7in) x 38 ring gauge. They have a closed head and open foot.

LONSDALES These fall between coronas and panetelas, longer than the first and thicker than the second. These are 17cm (6¾in) x 42 ring gauge.

PETIT CORONAS Suitable for a short smoke, these measure 12.7–14cm/5–5½ x 38–44 ring gauge.

Figuardos fall into the following main groups:

BELICOSOS These are small, pyramid-shaped cigars with a rounded head.

DIADEMAS These are the largest cigars, at more than 20.3cm/8in long. They can come either with an open or closed foot.

CULEBRAS These curious cigars consist of three panetelas braided or plaited together.

PERFECTOS These have two closed, rounded ends and a bulging middle.

PYRAMIDS These have a pointed, closed head and an open foot.

TORPEDOS These have a pointed head, a bulging middle and a closed foot.

LEFT: Arturo Fuente Double Corona

BASIC *PAREJOS* CIGARS AND THEIR SIZES		
CIGAR	LENGTH (CM/IN)	RING GAUGE
Double Corona	20cm/7⅞ in	49
Churchill	17.8cm/7in	47
Lonsdale	17cm/6¾ in	42
Panetela	17.8cm/7in	38
Corona	14cm/5½ in	42
Robusto	12.7cm/5in	50
Petit Corona	12.7cm/5in	42

ABOVE, LEFT TO RIGHT: *A range of cigar shapes including Don Ramos Petit Corona, Ashton Corona, Santa Damiana Torpedo, Cifuentes Belicoso and Canario d'Oro Lonsdale. Cigar manufacturers may use the same name to denote cigar sizes, although the exact dimensions will vary from brand to brand.*

A GUIDE TO HAVANA CIGAR SIZES

There are no fewer than 42 shapes and sizes of handmade Havana. They vary in length from 10–23cm (4–9in) – although some are smaller and some are larger – and in ring gauge from 26–52. Some names, like panetelas, are used to denote a particular size in a number of Havana brands. The problem is, the size is not necessarily the same in each brand, so you will need to check sizes. A Romeo Y Julieta Panetela, for example, is 11.8cm (4¾in) long with a 34 ring gauge while a Cohiba Panetela is 11.5cm (4 ½in) long with a 26 ring gauge.

Fortunately there is a comprehensive list of names available for the different shapes and sizes of Havana. It is the one used by the manufacturers of the cigars. In this list they have a *vitola de galera*, or factory name, for every type of Havana that is made.

Although a seasoned cigar smoker might have a favourite brand or size, many vary their choice

TOP: *A Hoyo de Monterrey Double Corona,*
a favourite with cigar aficionados.

ABOVE: *A Romeo Y Julieta Churchill.*
This was the first brand to name a cigar after
Sir Winston Churchill.

BELOW: *A Quintero Panetela,*
a mild brand of cigar made in Havana since 1940.

BOTTOM: *A Bolivar Lonsdale. This famous brand of Havana*
cigars has featured the portrait of the 19th-century
liberator, Simon Bolivar, since 1901.

NAMES, SIZES AND RING GAUGES FOR THE MOST POPULAR HAVANA CIGARS		
CIGAR	SIZE	RING GAUGE
Heavy Gauge		
Gran Corona	23cm/9in	47
Prominente	19.3cm/7⅝in	49
Julieta	17.8cm/7in	47
Piramide	15.5cm/6⅛in	52
Corona Gorda	14.25cm/5⅝in	46
Campana	12.7cm/5in	52
Hermosa No.4	12.7cm/5in	48
Robusto	12.4cm/4⅞in	50
Standard Gauge		
Dalia	17cm/6¾in	43
Cervantes	16.5cm/6½in	42
Corona Grande	15.5cm/6⅛in	42
Corona	14cm/5½in	42
Nacionale	14cm/5½in	40
Mareva	12.7cm/5in	42
Petit Cetro	12.7cm/5in	40
Belvedere	12.7cm/5in	39
Standard	12cm/4¾in	40
Franciscano	11.4cm/4½in	40
Minuto	11cm/4⅜in	42
Perla	10.2cm/4in	40
Slim Gauge		
Laguito No. 1	19cm/7½in	38
Ninfa	17.8cm/7in	33
Laguito No. 2	15.2cm/6in	38
Veguerito	12.7cm/5in	37
Seoane	12.7cm/5in	36
Carolina	12cm/4¾in	26
Cadete	11.4cm/4½in	36
Laguito No. 3	11.4cm/4½in	26
Chico	10.5cm/4⅛in	29
Entreacto	8cm/3⅛in	30

ABOVE LEFT: A Montecristo Corona, one of the most popular brands of Havana cigar.

ABOVE CENTRE: A Punch Robusto. This long-established brand of Havana was introduced in 1840 for the British market, and named after a popular magazine.

ABOVE RIGHT: A Romeo Y Julieta Petit Corona, a famous brand of Havana first introduced in 1875.

depending on the time of day. Smaller, lighter cigars are smoked before lunch and a big, full-bodied cigar after a heavy dinner. If time is short, experienced smokers go for the short but punchy robusto size.

THE COLOUR RANGE

As discussed earlier, cigar wrappers are normally classified according to seven basic colours, but they can come in a multitude of different shades. Generally, if you want a mild taste go for a light colour and for a full-bodied and sweeter flavour choose a darker colour.

The seven basic colours of wrapper leaf are the following:

DOUBLE CLARO These are light green in colour. The slightly sweet-tasting leaves are cured with heat to fix the chlorophyll in the leaf.

CLARO These wrapper leaves are taken from plants that are usually grown under shade tents. They are light tan in colour.

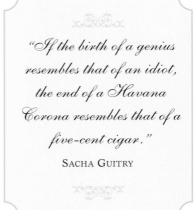

"If the birth of a genius resembles that of an idiot, the end of a Havana Corona resembles that of a five-cent cigar."

SACHA GUITRY

COLORADO CLARO OR CANDELA Often sun-grown, but sometimes grown under shade, these leaves are a light brown to brown in colour.

COLORADO These leaves are reddish-brown to brown in colour.

COLORADO MADURO Usually shade-grown with a medium-brown colour. The leaf possesses a subtle aroma and rich flavour.

MADURO These silky and oily leaves have a rich flavour and mild aroma. Extra time spent in the maturing process produces a rich, dark-brown leaf.

OSCURO The leaves are left on the plant for longer and the extra maturing and curing produces a leaf that is nearly black in colour. These are now rare in Cuba.

BELOW: H. Upmann Connoisseur No. 1 cigars. A good example of how cigars should all be an even colour.

BELOW: Macanudo Vintage Cabinet Selection – a brand of fine, mild cigars made in Jamaica.

ABOVE: Cigar colour can range from the dark brown maduro *at the top, reddish* colorados, *and various* claro *shades towards the bottom.*

RIGHT: Sorting Double Coronas by colour at the La Corona cigar factory.

During the final stages of production of Havana cigars, the *escogedor* (colour grader) sorts each type into at least 65 differing shades. A second colour grader then sorts them again so that, when placed in their cedarwood box, the very subtle difference in shades ranges from dark to light, left to right.

CIGAR QUALITY

andmade cigars fall into three categories and the price of a cigar increases with its quality.

PREMIUM CIGARS By far the most popular of the handmade cigars that are sold, these are made with long-leaf filler tobacco. It is worth noting that for some premium cigars the filler and bunching phases of manufacture are carried out by machine.

SUPER PREMIUM CIGARS These are made using specially selected tobaccos for the filler and the wrapper, and are subject to extra ageing.

VINTAGE CIGARS These top-of-the-range cigars are made exclusively from a single year's outstanding tobacco crop. As a result they are very expensive. Vintage cigars are highly collectable items, which can increase dramatically in value. In late 1996 a single box of 25 Trinidad Havanas was sold in Switzerland for $11,500 (£6,924) or $460 (£278) per cigar. In this case their rarity vastly increased their value. Trinidads are only made in one size (19cm/7½in x 38 ring gauge) and are made in the Cohiba factory in Havana. These cigars had never been sold to the public, as they were reserved for the sole use of Cuban Government officials and for diplomatic gifts. Plans were made in 1998 to make these available to the wider public on a commercial basis

LEFT: *A rare Trinidad. Originally made exclusively for Cuban Government officials and as gifts for foreign heads of state, they are likely to become available to the general public in the near future.*

RIGHT: *A mouth-watering display of premium handmade cigars, with some examples of the cigar-maker's tools – wooden moulds used for making "bunches" and the all-important* chaveta *or trimming blade.*

BUYING A CIGAR

A visit to one of the world's great cigar stores, such as Alfred Dunhill, Robert Lewis or Davidoff in London, La Civette in Paris, or Nat Sherman in New York, is more than just a normal shopping trip. For the cigar aficionado, it is to enter a world dedicated to one of life's main pleasures, full of cheering sights and smells, all of them redolent of calm and relaxation. The maturing room at Alfred Dunhill in London, dating from 1928, contains cabinets that can hold up to 30,000 cigars. In the same shop is a magnificent display cabinet dating from 1922, which contained Romeo Y Julieta cigars in all shapes and sizes. Although it is not the cheapest (due to high British import and tobacco taxes), London has a reputation as the best European city in which to buy handmade cigars, closely followed by

LEFT: An airtight container with five coronets, "selected and packed for campaigning" by Alfred Dunhill, for use by servicemen during World War II.

Geneva and Paris. The London branch of Alfred Dunhill, one of the establishments where Winston Churchill once bought his cigars, sells several hundred thousand handmade cigars a year, in over 200 brands and sizes.

Spain, where tobacco taxes are low, imports more cigars than any other European country, but the quality and authenticity of some of the Havana cigars is open to question. There are fine cigars to be bought in the best American shops, such as Nat Sherman, J. R. Cigars and Davidoff, all in New York, but, due to the continuing US trade embargo against Cuban imports, there are no Havanas available.

You can trust the well-known cigar shops, but it is worth being cautious when visiting unknown outlets. Be careful if you are tempted by sale bargains or duty-free cigars at airports: ensure that you are not being sold machine-made cigars, masquerading as famous handmade brands, by checking the stamps on the bottom of the box. If possible, try and look inside any box of cigars you are considering purchasing. If you can, buy cigars in boxes of 10 or 25. The larger packs are reputed to be of slightly better quality.

LEFT: The imposing entrance to Alfred Dunhill's maturing room in London.

RIGHT: Cedarwood cigar boxes are an art-form in themselves.

OPPOSITE: The walk-in humidor is a feature of many good quality cigar shops.

SELECTING A CIGAR

When selecting a cigar it is important to use all your senses. There are a number of rules that are worth following:

◆ Before you purchase a cigar, look inside the box. Check that the cigars are much the same colour. Contrary to myth, a dark Cuban cigar will not be any richer or full-bodied than one that is lighter in colour.

◆ Make sure that the wrapper leaves have a good sheen and that none are damaged or too heavily veined. The appearance of a wrapper will vary, depending on where the leaf was grown. Cuban wrapper leaves are silk-like, with a smooth surface; Cameroon wrappers have an oily and bumpy surface (called "tooth"); Connecticut wrappers have

> *"My father was a cigar smoker, and he really appreciated a fine cigar... He used to smoke Cuban cigars and drink Spanish wine. And he taught me about both things."*
>
> FIDEL CASTRO

more colour than an Ecuadorean one, for example, and have a good shine and slightly more "tooth".

◆ The cigar should not be brittle or dry. Feel a couple of cigars: they should be smooth and firm, but give slightly when you apply gentle pressure to them between finger and thumb. There should not be any crackling or rustling of leaves.

◆ Select a cigar to suit your needs. Novice smokers are advised to start with smaller cigars, for example the half corona size, normally around 12.7cm (5in) long, with a 42 ring gauge.

BELOW: Three famous Havana brands, from top to bottom: Punch, Bolivar and Romeo Y Julieta.

ABOVE: A slimline desktop humidor is convenient for storing a small number of cigars. Good cigar merchants will give advice on storing cigars and will probably stock humidors and other accessories.

RIGHT: A selection of cigars available at the Havana Club in Knightsbridge, London.

Smaller cigars are also better suited for daytime smoking. More experienced smokers will choose a big, full-bodied cigar, such as a Churchill or double corona, to smoke after a heavy meal.

Cigars continue to ferment and mature if they are correctly stored. However, fuller-bodied cigars age better than milder ones, which tend to lose their bouquet if stored too long. To assist in the ageing process, quality handmade cigars are not sold to the public for over a year after being imported. Some cigars can be left to mature for anything up to 10 years.

COUNTERFEIT CIGARS

In recent years Havana cigars have become prey to counterfeiters. Recent investigations by UK Havana importers, Hunters & Frankau, identified a large quantity of counterfeit Cohibas, Montecristos, Partagas and Hoyos, that were on their way to Britain. The shipment had originated in Havana, where the fake cigar trade has grown from a small-time operation aimed at tourists, into a larger scale criminal activity.

Cuba's government, supported by Habanos SA, its official cigar exporter, has mounted a major police operation to cut off supplies at source. However, the worldwide fine cigar boom, combined with continued economic problems at home, mean the rewards from this illicit trade are great. Tourists, predominantly in Havana, are bombarded constantly by street dealers, offering

> *"If a woman knows a man's preferences, including his preference in cigars, and if a man knows what a woman likes, they will be suitably armed to face one another."*
>
> COLETTE

cigars at very low prices, that have been made in small underground workshops. Often these counterfeit boxes bear all the attributes of a genuine box of Havanas, including the official warranty seal, the Habanos chevron, the hallmarks and the factory codes. It can be very difficult to tell the fakes from the real thing, even with an expert eye.

However, the truth becomes painfully obvious when the cigars are smoked, as they draw badly and taste awful. Sometimes they are even found to contain rolled up newspaper instead of tobacco! The counterfeiters concentrate their efforts on the most prestigious brands and only recently a selection of fakes was so convincing that they came close to being auctioned at Christie's in London.

To avoid such painful experiences, check that the box carries the five identification marks introduced to certify the authenticity of Havanas:

THE WARRANTY SEAL The Cuban Government Warranty Seal dates from a law enacted on 16 July 1912. This was one of the first steps taken to identify genuine Havana cigars.

THE THREE HALLMARKS These are found burnt into the underside of the cigar box:

◆ *Habanos S.A.* The name of the Cuban company which has exported Havanas since 1 October 1994. Boxes made between 1985 and 1994 have the logo of Cubatabaco instead.

◆ *Hecho en Cuba* This was added in 1960, and replaced the English label that was used before, with the words "Made in Havana – Cuba".

◆ *Totalmente a Mano* Literally "Totally by Hand", this has appeared from 1989 onwards, and establishes that the cigars are fully handmade in the traditional Cuban manner. This is an important point, as EU rules permit partly machine-made cigars to be described as "handmade".

BELOW: The Habanos chevron has been printed across the corner of all boxes of genuine Havanas since 1994.

If the Warranty Seal, or any one of these marks is missing, you can be sure that the box does not contain genuine handmade Havana cigars.

THE HABANOS CHEVRON The most recent addition is found on boxes manufactured from 1994 onwards. This is the word "Habanos" printed in a chevron across the corner of the box.

The soundest advice to avoid buying fake Havanas, is to avoid purchasing cigars that are offered on the cheap, and to always buy them from a reputable tobacco merchant.

ABOVE LEFT AND LEFT: The Cuban Government's warranty for cigars exported from Havana dates back to 16 July 1912. Even over 85 years ago, this warranty was deemed necessary, so that genuine Havanas could be distinguished from fakes.

RIGHT: Always check the underside of a cigar box – the three hallmarks that are burnt into all boxes of genuine Havana cigars are: Habanos S.A., Hecho en Cuba and Totalmente a Mano.

CIGAR PACKAGING

Several centuries ago, cigars used to be packed in bundles and covered with pig's bladders. Vanilla pods were placed next to the cigars to help enhance the smell. Later this practice was abandoned in favour of large cedar chests, holding thousands of cigars. The use of cedarwood helps to arrest the drying-out process and enhances maturing. Then, in 1830, the bank of H. Upmann decided to send cigars to its senior staff in London, packed in the small cedarwood boxes that are still familiar today. The boxes were stamped with the bank's symbol. The company of H. Upmann eventually decided to diversify into the cigar business and other Havana cigar-makers soon copied their form of packaging.

Branding cigars started at around the same time. As the cigar industry grew, simple marketing became necessary, so that the manufacturers could differentiate between their products. In 1837 a Cuban cigar-maker, Ramon Allones, started applying a colourful label to his boxes of handmade cigars. Other manufacturers followed suit and soon the illustrated labels, clearly identifying each brand, were being

ABOVE: The firm of H. Upmann was the first to package its cigars into small cedarwood boxes.

LEFT: A box of 25 Valdrych 1904 premium cigars, handmade in the Dominican Republic.

RIGHT: A fine selection of handmade Havanas. Colourful labels were first applied to the cedarwood boxes in 1837.

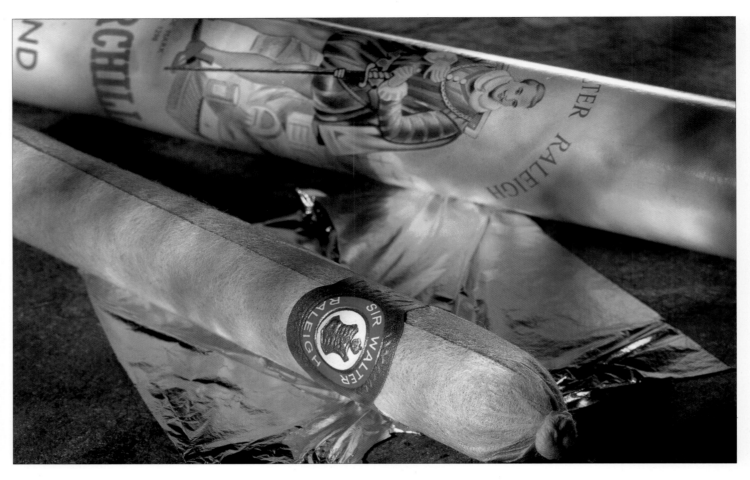

ABOVE: A romanticized version of Sir Walter Raleigh – even aluminium cigar tubes can be a work of art!

"To smoke is human; to smoke cigars is divine."

ANON

applied to the front and inside of the box lids. These boxes, with their colourful labels, soon became attractive objects in their own right. From 1912, Havana cigars were also sealed with a green label to guarantee authenticity. Soon this custom was copied by makers of handmade cigars in other

countries and the practice still continues today. Certain premium cigars are packed in varnished cedar boxes. At the very top end of the market, some cigars are packed in expensive cedar cabinet boxes, fitted with brass hinges and screws.

It is worth repeating that the bottom of all boxes of cigars must be examined carefully. For Cuban cigars made before 1961, just after the Revolution, the underside of the box will

LEFT: A cedarwood box of Fox's own brand of handmade cigars, made exclusively for J.J. Fox's of London, England, whose past clients include Sir Winston Churchill and Oscar Wilde.

RIGHT: A fine box of 25 H. Upmann Gran Coronas, the label proudly displaying the many gold medals won by the company in the 19th century.

RIGHT: The Cuban warranty seal will be clearly displayed on every box of authentic Havanas.

be stamped "Made in Havana – Cuba". Those made subsequently will be stamped "Hecho en Cuba". From 1985–1994 these cigar boxes were also stamped with the code of the factory and the symbol of Cubatabaco, the state tobacco monopoly. The Havana factory codes use the initials of the post-revolutionary factory name: FPG – Francisco Perez German (Partagas); FR – Fernando Roig (La Corona); HM – Heroes del Moncada (El Rey del Mundo); JM – Jose Marti (H. Upmann); BM – Briones Montoto (Romeo Y Julieta); EL – El Laguito.

From 1994 up to the present this has been replaced with "Habanos S.A.". Finally, to ensure that Havana cigars are handmade, the underside of the box must be stamped "Totalmente a Mano" – this stamp has been applied since 1989. The same goes for other non-Havana cigar brands which should be stamped "Handmade".

LEFT: Arturo Fuente is a much sought-after brand of handmade Dominican cigars.

The stamp "Envuelto a Mano", which you will find on boxes of non-Havana cigars, means they are hand-packed only. The designation "Hand-rolled" or "Hand-finished", only means that the wrappers on the cigars are applied by hand, with the rest of the process carried out by machines. The most expensive vintage cigars are stamped with the year of the tobacco crop.

BELOW: Hermetically sealed jars, complete with leather carrying handle, containing H. Upmann "nature-seasoned" cigars.

THE CIGAR BOX LABEL

Cigar box labels are elaborate affairs, often involving different designs on the inside and outside of the box, with a third design on the lining paper. Many of these are highly collectible, designed in full colour, with ornate printing techniques, and often gold embossing. Depending on the rarity of the label, they can fetch reasonably high prices (although not yet comparable with the price of vintage cigars themselves). Early editions, featuring cigar-smoking

monarchs or other personalities, are highly sought after. But whether the labels are valuable or not, there is no denying their visual appeal.

ABOVE, LEFT, AND BELOW. Labels might include elaborate gold-embossed designs, or well-known faces such as this label featuring Edward VII.

ABOVE AND LEFT: *Classical and allegorical depictions and heraldic designs were popular motifs on Cuban labels.*

BELOW: *This label was designed for the US market. Even in the early days, cigar manufacturers put a great deal of thought into marketing their product.*

THE CIGAR BAND

In the middle of the 19th century wealthy smokers of Havanas found that stains from cigars were spoiling their white gloves. To solve this problem, the cigar band was introduced by Gustave Bock, a Dutch manufacturer of Havana cigars. Originally the band consisted of plain white paper which stopped the staining. However, the cigar band soon

RIGHT AND BELOW: Otto von Bismarck, the 19th-century Chancellor of Germany (right) and King Edward VII (below), both smokers of Havanas, had personalized labels applied to their favourite cigars.

became very ornate, following the introduction of boxes and decorative labels as a means of differentiating between brands. Some wealthy Havana smokers, such as King Edward VII and Bismarck, even had their own personalized bands bearing their portraits. These early bands are now extremely valuable collector's items. In recent years the designs have become simpler. Some brands, such as the Cuban Romeo Y Julieta, use more than one band design. The brand's Churchill size, for example, has a simple gold band, whereas all its other sizes have red ones. The bands on non-Havana cigars with Cuban brand names usually use a similar design, but with subtle differences. For example, the Cuban-made Romeo Y Julieta cigar band is printed "Romeo Y Julieta – Rodriguez Arguelles Y Ca. – Habana", while the Dominican-made version is printed "Romeo Y Julieta – Medallas de Oro – Dominican".

The origins of the names of cigar brands vary greatly. Punch, created for the British market in 1904,

"Tobacco is the plant that converts thoughts into dreams."
VICTOR HUGO

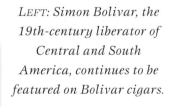

LEFT: Simon Bolivar, the 19th-century liberator of Central and South America, continues to be featured on Bolivar cigars.

RIGHT: Montecristo cigars, with their white-and-brown band, outsell all other Havanas.

OPPOSITE: A box of vintage cigars with bands intact.

ABOVE: Distinctive designs are a feature of cigar bands, many of which have not changed since the 19th century.

Like stamp collecting, cigar bands can be an interesting and worthwhile hobby, with some of the older versions being much sought after and increasing in value. Even those that are not highly collectible can be appreciated as works of art in their own right.

ABOVE: Sorting finished handmade cigars into brand types in a Havana cigar factory.

was named after the popular magazine of the time; Bolivar is named after Simon Bolivar, "The Liberator", who freed Central and South America from the Spanish yoke, and Romeo Y Julieta was one of the favourite stories of Havana cigar rollers.

In applying the band, the bander (or *anilladora* in Cuba) takes the cigars out of the box to dress them. Under no circumstances may the bander alter the order or face chosen by the colour grader (or *escogedor* in Cuba). The band is placed on the selected face of each cigar, which is then carefully replaced in the prescribed order of colour tones.

Whether a cigar is smoked with the band on or off is entirely up to the smoker, although in Britain it was once considered "bad form" to show the brand you were smoking. If you prefer to smoke a cigar without its band, do not remove it until you have smoked it for a couple of minutes. That way, the heat will melt the vegetable gum (glue) on the band and allow its removal without damaging the wrapper leaf.

LEFT AND BELOW: The famous Punch brand of Havana cigars, one of the oldest still made, was introduced for the British market in 1840 by J. Valle & Co. The cigar-smoking character of Mr Punch and his faithful dog, seen gracing this 1915 cover of the famous humorous magazine, still appear on the present-day cigar boxes. Although production of this large range of cigars still continues in Havana, in recent years a separate range with the same name has also been made in Honduras.

ABOVE: *A selection of beautifully designed cigar bands, from Cuba and the Dominican Republic, show the importance that manufacturers place on individualism.*

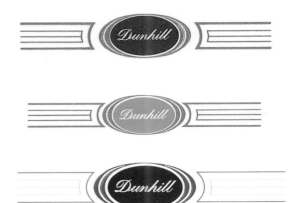

ABOVE: *Gustave Bock was the first manufacturer to use decorated cigar bands. Henry Clay cigars were named after a 19th-century American senator.*

LEFT: *Modern bands used on Dunhill cigars.*

CIGAR EPHEMERA

Since cigar-smoking became popular in the 19th century, there has grown a host of accompanying accoutrements and pieces of ephemera associated with the pleasure of smoking. For the solitary smoker, a smoking cap was *de rigeur;* as cheroots became popular, so increasingly lavish holders were designed, made from ivory and amber. Greetings cards proliferated, referring to the joys of smoking. During World War I, postcards were issued by the Red Cross: for every one sold a cigar was sent to a soldier in the trenches.

ABOVE: A postcard issued by the Red Cross, proceeds of the sale of which would allow a cigar to be sent to soldiers.

BELOW: A pattern for a smoking cap, originally published in a popular women's magazine.

RIGHT: Cigar paraphernalia, including playing cards, a smoking cap and a cheroot holder.

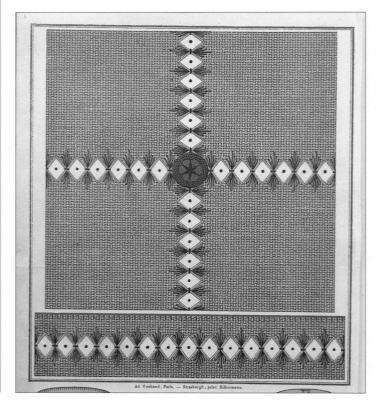

Ad. Goubaud, Paris. — Strasburgh, print. Silbermann.

CUTTING THE CAP

When you are preparing to smoke a cigar the first thing to do is to cut the cap. This allows the cigar to draw properly. All handmade cigars, and some of the more expensive machine-made brands, have a closed end covered by a cap. Cutting this cap is an essential part of the cigar-smoking ritual. It needs to be cleanly and levelly cut between 1.5–3mm (⅛–¹⁄₁₆in) above the base of the cap. The method used for this is a matter of personal choice. You can use a special cigar cutter or just as easily use a sharp knife. You can even use your fingernails, if you are skilled enough, in which case you simply pinch off the top of the cap.

There are a large number of cigar cutters on the market, ranging from small, cheap, easily portable plastic guillotines to much fancier versions: small, decorative works of art in their own right made of steel, silver, gold and enamel. Guillotines come in single- or double-bladed versions, but the latter is the best tool for the job. A fine example

ABOVE: Professionals in cigar shops have mastered the art of using cigar scissors to cut the cap.

BELOW: A beautifully made Hillwood guillotine cutter for the real cigar aficionado.

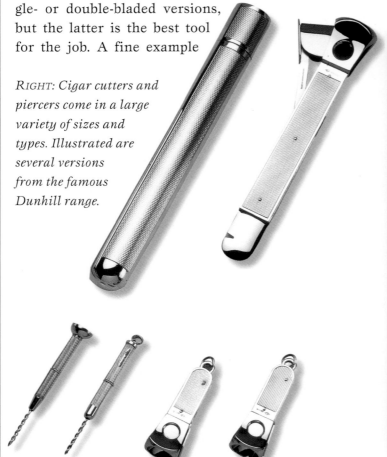

RIGHT: Cigar cutters and piercers come in a large variety of sizes and types. Illustrated are several versions from the famous Dunhill range.

ABOVE: The correct way to use cigar scissors. Whatever method you use, cutting the cap is an essential part of the cigar smoking ritual.

with a hollow tip that is turned in a circular manner. Both these methods are no longer considered very satisfactory, as the single hole can concentrate unpleasant acids and juices on the tongue. Some smokers actually prefer to use a razor blade or even bite off the cap with their teeth. However, the practice of spitting out the cap is not considered good manners! Whichever type of cutter you use it should always be kept sharp. A dull blade will tear the tobacco, leaving a jagged cut and ruining the cigar.

Cigar cutters are now collector's items and can come in a huge variety of shapes and sizes. Often dating from the mid-19th century they can be worth anything from $165 (£100) to over $3,300 (£2000), depending on condition, rarity and material used. They normally come in three basic types: pocket-sized cutters or punchers, cigar scissors and large cigar store models, or combination sets.

of a good-quality single-bladed guillotine is the Paul Garmirian Cigar Cutter, while Davidoff's Zino Cigar Cutter is an excellent example of a two-bladed version.

Whichever type of cutter you use, make sure it is sharp. Do not pierce the cap, as some smokers still do – it will make the cigar overheat by compressing the vital filler leaves and can completely ruin the flavour. Make sure you cut the cap carefully. If you don't, it will seriously impair your enjoyment of your cigar. If the cut isn't level, the cigar's draw will be affected, and it will heat up unevenly. Ensure that the wrapper leaf is not damaged and, whatever you do, don't cut below the level of the cap, since this will certainly damage the wrapper.

There are also a number of cigar scissors available, that are used by professionals in cigar shops, but they take a little mastering. Another type of popular cutter makes a wedge shape in the cap, which exposes a greater surface area of the filler bunch.

There are two other types of cutter: a piercer and a bull's-eye. They both pierce a hole in the end of a cigar without removing the cap. The bull's-eye uses a cutter

BELOW: A fine selection of beautifully made pocket-sized cigar guillotine cutters. Whether you choose cutters or scissors, always ensure the blade is sharp for best results.

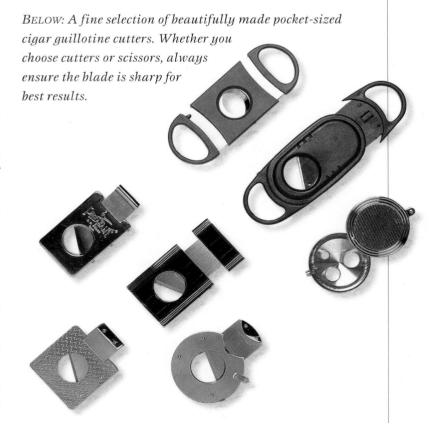

LIGHTING A CIGAR

Lighting a cigar is not merely a matter of applying a flame to it: a well and carefully lit cigar will always be more enjoyable than one that is badly lit. It doesn't matter whether you use a match or a lighter, although a lighter should burn butane rather than gas. Some specialist cigar lighters actually have two flames. Most major cigar shops supply long, slow-burning cedarwood matches that do the job very well, but you can just as easily use an ordinary match, providing it is not too sulphurous or waxy. However, do not use a cardboard match impregnated with chemicals. The important thing is that the cigar is evenly lit and that its flavour is not contaminated by the source of the flame.

BELOW: Long, slow-burning matches can be purchased from most cigar stores.

ABOVE: After holding the end of the cigar in the flame, draw slowly, while gradually rotating it.

RIGHT: A specialist cigar lighter, such as this Dunhill model, burns butane rather than gas.

Ensure that you have a decent-sized flame. Hold the cigar horizontally, with the flame just touching it. Then, slowly revolve the cigar until the end is evenly charred. Hold the flame about 12.5mm (½in) away from the end and draw slowly on the cigar, keeping it horizontal and continuing to turn it. The end of the cigar should now ignite. Gently blow on the glowing end to ensure that it is burning evenly. This is important, otherwise one side will burn faster than the other, which will affect the flavour and draw of the cigar, and increase the likelihood of it going out. Never inhale cigar smoke, just hold the smoke in your mouth and then exhale. It is worth noting that older, well-matured cigars burn more easily than less mature ones. Warming the length of a cigar to remove excess gum (glue) is no longer necessary, as modern cigars only use flavourless and odourless vegetable gum.

The key to maximum enjoyment of a cigar, once it has been properly lit, is to avoid overheating the filler.

Smoke slowly, without puffing too much on the cigar, or dragging on it. Puffing should occur at intervals of about one minute, in order to keep the cigar alight. A gently puffed cigar should last about 45 minutes, depending on its size, but try and avoid the head of the cigar becoming wet with saliva. This will not only affect the taste, but also looks rather unsightly. If the cigar goes out, tap away any ash, re-light it as before, and blow through it to remove any stale smoke before you start smoking again. A re-lit cigar may smoke stronger, but this is sometimes better than nothing! However, relighting a cigar after more than a couple of hours will almost certainly mean that its flavour is significantly impaired. You do not need to tap a cigar to remove its ash, as it will fall off when ready. The time to discard a cigar is usually when about one-third of the length remains: this is when it starts to produce hot smoke and a strong after-taste. Do not stub out a discarded cigar – this will leave an unpleasant odour – but leave it in the ashtray and it will eventually go out. A good cigar will leave a residue of long, firm ash. Finally, remove cigar stubs before the room fills with stale smoke.

BELOW: A cigar that goes out should be re-lit within a couple of hours.

STORING AND CARING FOR CIGARS

Cigars are a natural, organic product and need to be stored just as carefully as you might store food or wine. The air-conditioned or centrally heated home is their biggest enemy. Cigars should be kept in a humid environment, avoiding too much heat or cold, ideally at about 70%–72% humidity, and within a temperature range of 18°–21°C (64°–70°F). Kept within these temperature limits, cigars will not become dry or brittle, and the environment will assist in the ageing process. Cigars should never be stored in direct sunlight or exposed to sea air. Really serious cigar smokers entrust their reserve of cigars to the care of their favourite cigar

> *"My boy, smoking is one of the greatest and cheapest enjoyments in life, and if you decide in advance not to smoke, I can only feel sorry for you."*
>
> SIGMUND FREUD

merchant, while others have specially adapted cellars or closets in their homes for the purpose. However, there are practical measures that anybody can take.

HANDMADE CIGARS

The key to storing cigars properly is to make sure they don't dry out. Therefore, keep them contained in their cedar boxes, which are designed to assist in cigar preservation, in an airtight cupboard or box, well away from any heat sources, in a cool part

BELOW: Although expensive, a wooden humidor like this, if well cared for, should prove to be a good investment.

ABOVE: Cedarwood-lined cigar tubes are convenient to carry for the cigar-smoking traveller.

of the house. It also helps to keep a damp sponge in the cupboard and to check from time to time that it is still moist. If you only possess a few cigars, put the box that they are packed in into a sealed polythene (plastic) bag, having sprayed the inside of the bag with a small amount of water. Take great care to avoid contact between cigars and water, as this will create mould. One method of storing cigars, that is frowned upon by some experts, is to place them in an airtight polythene bag, taking care to suck out all air beforehand, and keep them in the vegetable compartment of a refrigerator. If this method is adopted, then the cigars should be taken out of the refrigerator at least 30 minutes before smoking. This enables the

RIGHT: Pocket cases are handy for holding a few favourite cigars.

temperature of the cigar to increase to the room temperature. Unwrapped cigars should never be stored in the refrigerator.

Cigars in aluminium tubes lined with cedar, though very convenient to carry, can sometimes become rather dry, as the tubes are not completely airtight. Ideally, cigars should be stored in aluminium tubes for no more than three days. If cigars are wrapped in cellophane, they are best left in their wrappings, unless you have a trustworthy humidor into which you can transfer them swiftly. Major cigar shops also sell small humidifiers which can be placed in the cedar box. Also available are telescopic pocket cigar cases, usually made of leather, that can be adjusted to accommodate a small number of cigars.

Alternatively you could buy a humidor, a special container designed to keep cigars moist. The types of humidor available are extensive, and range from small ones, made of wood or leather and intended to be used when travelling, to major pieces of furniture. Quality humidors are made by manufacturers such as Davidoff, Dunhill, Savinelli and Diamond Crown. The larger humidors, essentially large pieces of furniture, are made by such manufacturers as Vinotemp, Prendegast and Kretman-Thelan.

LEFT: Polythene (plastic) bags are a cheap alternative for storing cigars and keeping them fresh.

BELOW: Valdrych 1904 premium cigars stored in a small Dunhill humidor.

Humidors are traditionally made of wood such as walnut, mahogany and rosewood, some costing large amounts of money, but there are also cheaper plastic models now available. It is important that a humidor is well-made, with perfectly squared and fitted seams, unvarnished inside, contains a hygrometer to measure the humidity level and, above all, that it has a heavy, tight-fitting lid. Other important features should include a tray for storing cigars at different levels, and air slots in the box sides to allow the unit to breathe, thus preventing warping and separation of the box's veneer. Most humidors have humidification systems that require water. It is recommended that distilled water is used, as tap water contains damaging chemicals. The price of a humidor can range from about $205 (£125) to many thousands for the larger models. An antique humidor once owned by President John F. Kennedy was recently sold at auction in the USA for over $580,000 (£350,000). Remember, that whatever humidification system it uses, from a sponge to chemicals, you cannot just go away and forget it: the system will need to be checked and topped up from time to time. For the cigar-smoking traveller, some tobacco companies make briefcases that contain small humidifiers.

Because of the importance of proper storage, when you buy cigars from anywhere other than a major cigar shop, do check that they aren't just kept on a shelf or in a cupboard: you are pretty sure to be disappointed in your purchase if they are.

LEFT: A small, cedarwood-lined cigar box with a simple built-in humidor, fitted with a meter to measure humidity.

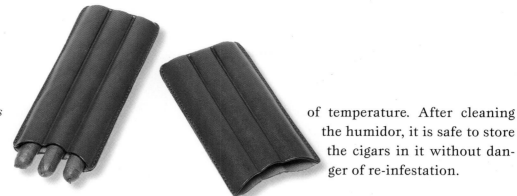

RIGHT: Three Bolivar Royal Coronas fit snugly into this telescoping leather pocket cigar case.

of temperature. After cleaning the humidor, it is safe to store the cigars in it without danger of re-infestation.

STORING MACHINE-MADE CIGARS

European dry cigars should be stored at a humidity of 11–13%, ideally in a place where humidity and temperature do not vary much. The recommended temperature is 18–20°C (64–68°F) with extremes of 12–25°C (54–77°F). Severe cold, excessive heat and sudden changes in temperature should be avoided. However, the shelf-life of the Dutch dry cigar is practically unlimited under normal climatic and handling conditions. A humidor is not required and several years' storage does not affect their quality, provided they are not humidified, and not moved about frequently. The length of time will, in fact, improve and enhance their flavour.

CIGAR PESTS AND DISEASES

MOULD This can occur if cigars come into contact with water or the humidor is not kept clean. Destroy all of the mouldy cigars and ensure that the humidor is thoroughly cleaned and dried before restocking it. A white bloom on a cigar's wrapper is caused by oils being given off during the ageing process. If this occurs, then the bloom can be wiped gently from the cigar with a soft cloth.

TOBACCO WORM Sometimes the eggs of the tobacco worm can escape the fumigating process during manufacture. The eggs, lying dormant inside the cigar, hatch out in the tropical climate of a humidor. The first signs of tobacco worm are small bore holes in the cigars. If this occurs, then destroy all the affected cigars immediately, and place the undamaged ones in a sealed plastic or polythene bag in a freezer for several days. This will guarantee to stop any further reinfestation. Remove the cigars, still in their plastic bag, from the freezer to a refrigerator for a few more days, before returning them to room temperature. This ensures that the wrapper does not split due to sudden changes

RESUSCITATING CIGARS

With a bit of luck, a dried-out cigar can be resuscitated. A simple method is to place the opened box of cigars, with a moist sponge, into a partially closed polythene (plastic) bag. Move the cigars around in the box over a period of several days and after about two or three weeks they should be ready to smoke. Unfortunately, this method does not return the lost bouquet to the cigar. Another simple and quicker method of bringing life back to dry cigars is to moisten the underside of the box and place it in an airtight polythene bag for a few days.

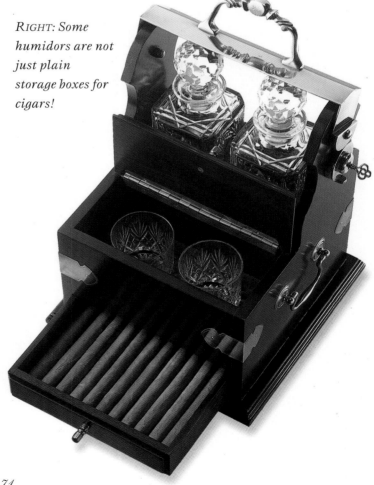

RIGHT: Some humidors are not just plain storage boxes for cigars!

CIGAR SOCIETY

As a cigar smoker you will be in good company. Certainly the best-known smoker of cigars is Winston Churchill, who during his long lifetime is reputed to have smoked about 250,000 cigars. He was introduced to cigar smoking in his early 20s, when he served in Cuba during the short Spanish-American war of 1898. He is reckoned to have smoked about 4,000 cigars a year, selected from his specially constructed humidified room, until his death at the grand old age of 91. His favourite cigar was a 47 ring gauge double corona maduro, a type which is now named after him. During World War II, at the height of the German airforce blitz on London, the premises of his

"And a woman is only a woman, but a good cigar is a Smoke."

RUDYARD KIPLING, THE BETROTHED, 1886

favourite supplier of cigars, Alfred Dunhill, were damaged by bombing. The store manager's first task, in the early hours of the morning, was to telephone Churchill to reassure him that his cigars were unharmed.

Amongst US presidents, John F. Kennedy was a well-known cigar aficionado. The story goes that just before the US embargo on Cuba in 1962, he requested a trusted aide to collect a large personal supply of his favourite

LEFT AND FAR LEFT: Oscar Wilde, the Irish writer and celebrity, was a keen cigar smoker. Here he is pictured in a flamboyant stance, gloves in one hand and cigar in the other. His account book at J.J. Fox's of Piccadilly in London, from the 1890s, is pictured on the far left.

RIGHT: General (later President) Ulysses S. Grant, here pictured in Vanity Fair magazine, was a cigar chain-smoker.

RIGHT: *Communist Fidel Castro, pictured here chomping a cigar during the first days of January 1959, when the Batista regime was ousted, was responsible for the nationalization of Cuban cigar companies.*

BELOW: *Charlie Chaplin, here pictured in the 1931 film* City Lights, *was another well-known lover of cigars.*

gauge) and Valdrych Quisqueya Real (25.4cm/10in x 50 ring gauge) are probably the largest cigars available.

Other well-known cigar-smoking characters, past and present, include US President Ulysses S. Grant, Mark Twain, Otto von Bismarck, Rudyard Kipling, King Edward VII, US President John Quincy Adams, Albert Einstein, Sigmund Freud, Somerset Maugham, Babe Ruth, George Burns (who smoked cigars until he was 100 years old), Charlie Chaplin, Groucho Marx,

BELOW: *German Chancellor Otto von Bismarck is seen during treaty negotiations in 1871, giving Favre his views on cigars.*

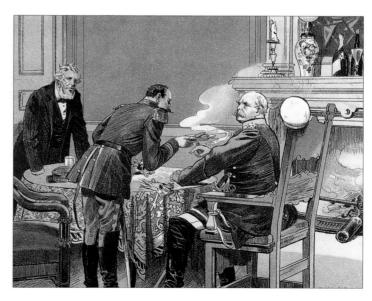

Havanas, H. Upmann Petit Coronas. King Farouk of Egypt, deposed by Nasser in 1952, once smoked a specially made cigar called the Visible Immenso, which was 46cm (18in) long with a 47 ring gauge. Probably one of the largest cigars ever made can still be seen in the London store of Davidoff. It is almost 90cm (36in) long, with a ring gauge of 96. Currently, the Royal Jamaica Ten Downing Street (27cm/10½in x 51 ring

LEFT: *The French novelist Colette was a committed cigar smoker and apparently smoked them in bed.*

RIGHT: *Bonnie Parker, of Bonnie and Clyde notoriety, reinforces the glamorous, rebellious image of the female cigar smoker that persists to this day.*

Edward G. Robinson, Fidel Castro, Bill Cosby, Lucille Ball, Arnold Schwarzenegger, Madonna, Whoopi Goldberg and Bette Midler – the list is endless!

Since the 19th century, cigars have been a strong symbol of power, prestige and privilege – however, this tends to refer more to male smokers than their female counterparts. In recent years, many young, glamorous female celebrities have gained attention from smoking cigars. Although today this creates an image both of power and, to an extent, rebellion (in light of well-known health risks), cigar smoking among women has been practised since the 19th century. Many female cigar smokers moved in artistic or bohemian circles, and artistic images of women smoking symbolized several ideas in the 19th and early 20th centuries, ranging from women's increasing freedom, to an indication that the smoker lived outside society's rules.

In recent years, the popularity of cigars in the US has soared. In the wake of this, exclusive cigar clubs

ABOVE AND RIGHT: *Cigar smoking among women had a certain risqué allure in the 19th century. The advertisement above shows leisured ladies enjoying Havanas. Compare this to* Chica in a Bar *by Ramon Casas I Carbo (right), in which the cigar-smoking subject is portrayed in a very different light.*

have sprung up in many major cities throughout the country, although the recent ban on smoking in public places, imposed by several state authorities, including California, will severely restrict or make illegal these establishments.

CIGAR DIRECTORY

Cigar smoking is currently enjoying a renaissance. The demand for premium handmade cigars often exceeds supply, and cigar aficionados are prepared to pay large amounts of money to luxuriate in their aroma and taste. Currently, the largest proportion of handmade cigars is only available in the United States – by far the largest consumer, thanks partly to the stars of cinema and TV, who have widely publicized their enjoyment of the habit. Out of a total of over 1,000 brands, about 90 % can be bought in the United States, while only 15 % can be purchased in the UK. However, this small percentage does include the availability of most Havana cigars, due to the continuing US embargo on Cuban goods.

BRAZIL

Brazil is not a major producer of hand-made cigars, and probably the most well-known Brazilian cigar is the medium-flavoured Suerdieck range, which produces a medium-quality cigar. These small, ring-gauge cigars, mainly machine-made, but some still handmade, all use home-grown tobacco for wrapper, filler and binder. The recently introduced Don Pepe range of good-quality hand-made cigars, also produced by Suerdieck, is made with a Brazilian-grown Sumatra leaf and has proved popular with North American smokers. Zino is another range of small machine-made cigars produced in Brazil. Brazilian cigar tobacco is dark and rich with a slight, sweet taste.

ABOVE: *Brazil's home-grown tobacco is used in its popular Don Pepe range of cigars, amongst others.*

BELOW: *The coastline of Rio de Janeiro.*

CANONERO

This Brazilian brand combines Connecticut Shade wrappers with Brazilian binders and fillers, resulting in a mild cigar of excellent quality. The brand has been available since 1995, and special gift boxes for the Lonsdale and Robusto sizes have recently been introduced.

NAME	SIZE	RING GAUGE
Double Corona	19cm/7½in	50
Churchill	17.8cm/7in	46
Lonsdale	16.5cm/6½in	42
Rothschild	14cm/5½in	50
Corona	14cm/5½in	42
Robusto	12.7cm/5in	52
Potra	10.8cm/4¼in	38

RIGHT AND FAR RIGHT: Canonero cigars are available in different-sized boxes, providing a unique, mild flavour at an excellent price.

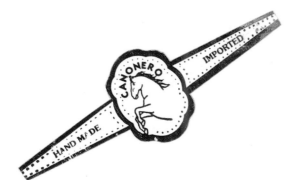

DON PEPE

Introduced by Suerdieck in 1994, this range of good-quality, mild- to medium-flavoured, handmade cigars has proved to be a success in the USA. They are made with Brazilian wrappers from tobacco plants grown from Sumatran seed.

NAME	SIZE	RING GAUGE
Double Corona	19cm/7½in	52
Churchill	17.8cm/7in	48
Petit Lonsdale	15.2cm/6in	40
Slim Panetela	13.3cm/5¼in	26
Robusto	12.7cm/5in	52
Half Corona	10.8cm/4¼in	34

SUERDIECK

A range of mild- to medium-flavoured, handmade cigars that is of medium quality and construction. The cigars are made with all-Brazilian fillers, binders and wrappers. Suerdieck also manufactures machine-made cigars.

NAME	SIZE	RING GAUGE
Nips	15.2cm/6in	32
Fiesta	15.2cm/6in	30
Valencia	15.2cm/6in	30
Caballero	15.2cm/6in	30
Finos	14.6cm/5¾in	46
Corona Brasil de Luxe	14cm/5½in	42
Brasilia	14cm/5½in	30
Mata Fina Especial	13.3cm/5¼in	42
Viajantes	12.7cm/5in	40
Mandarin Pai	12.7cm/5in	38
PREMIUM SERIES		
Panetela Fina	14.6cm/5¾in	36
Corona Brasil Luxo	14cm/5½in	45
Corona Imperial Luxo	14cm/5½in	45
Mata Fina Especial	13.3cm/5¼in	42

CAMEROON

The African republic of Cameroon, a former French colony, is not a producer of cigars, but its tropical climate helps it to grow some of the most sought after wrappers in the world, renowned for their rich, dark appearance. The leaf has a spicy flavour and a sharp smell. Cameroon wrappers are used in the making of many well-known,

premium, handmade cigar brands, including favourites like Cuesta-Rey, Don Diego, Montecruz, Nat Sherman, Partagas, Dominican Ramon Allones, Dominican Romeo Y Julieta and Royal Jamaica.

ABOVE: Nat Sherman cigars (and the Nat Sherman shop) are renowned throughout the world. Cameroon wrappers are used in many of the brand's cigars.

ECUADOR

The South American country of Ecuador, with its coastal tropical climate, now grows a good-quality, Sumatran-seed wrapper, known as Ecuador/Connecticut. The high temperatures, combined with a humid climate and fertile soil, make it an ideal location for growing fine tobacco plants. Ecuadorean cigars are generally mild, with a good flavour.

Tobacco grown in Ecuador is widely exported and used in

the making of many handmade cigar brands, including Bances, Cuba Aliados, the Honduran brand of El Rey del Mundo, Honduran La Gloria Cubana, Honduran Punch (not to be confused with the celebrated Cuban brand of the same name), Hoyo de Monterrey and Sosa.

LEFT AND ABOVE: Ecuadorean tobacco is used in many fine brands, including Bances and the Honduran Punch.

THE CANARY ISLANDS

The Canary Islands are a chain of seven volcanic islands, owned by Spain, in the Atlantic Ocean. As well as being a well-known vacation destination, they are the agricultural centre for superior-quality tobaccos. The north-

west coastline of Africa lies a short distance to the east and the dominant air and water flow is from the north-east to the south-west. Locally grown filler and binder tobacco is used for many of the cigars made on the islands, but additional filler and binder material is imported from Brazil, the Dominican Republic and Indonesia. Wrappers are imported from many sources, including the USA.

The tobacco industry in the Canaries benefited from the 1959 Cuban revolution. The American Consolidated Cigar Corporation formed Cuban Cigar Brands with Pepe Garcia, and set up a factory in the Canary Islands, producing premium, handmade cigars. Garcia was one of the major Cuban manufacturers whose factory was nationalized by the Castro regime. He and another Cuban manufacturer, the Menendez family, started making Montecruz cigars using Dominican and Brazilian fillers and Cameroon wrappers. Since the mid-1970s, Menendez has moved the manufacture of these cigars to the Dominican Republic.

ABOVE AND BELOW: The climate of the Canary Islands makes it ideal for growing superior-quality tobacco leaf.

DUNHILL

Today, the well-known English company of Alfred Dunhill markets a small range of cigars that is made in Tenerife by CITA. The cigars have a filler from Brazil, Santa Domingo *Piloto Seco* and Santa Domingo *Piloto Ligero*, a Santa Domingo binder and a Connecticut wrapper. The five sizes have a mild to medium flavour with a touch of sweetness, and are currently only available in the USA. This range can be identified by its red bands. Dominican-made Aged Dunhill cigars have a blue band.

NAME	SIZE	RING GAUGE
Lonsdale Grande	19cm/7½in	42
Corona Grande	16.5cm/6½in	43
Panetela	15.2cm/6in	30
Corona Extra	14cm/5½in	50
Corona	14cm/5½in	43

BELOW: *Colour-graders sort Dunhill cigars prior to labelling.*

CUBA

The original inhabitants of Cuba, the Taino Indians, smoked crude "cigars" made from rolled tobacco leaf. Today, this Caribbean island is widely acknowledged as still being the producer of the finest cigars and cigar tobacco in the world.

The finest tobacco is produced in the Vuelta Abajo province of Pinar del Rio, near the towns of San Luis and San Juan y Martinez. The climatic and soil conditions in this part of Cuba are perfect for the growing of tobacco, and it is the only region that produces all the leaves needed to blend a cigar. The province's annual rainfall of over 152cm (60in) a year is one of the highest on the island. The main growing period is between November and February, the dry season, when temperatures average 27°C (81°F) with eight hours of sunshine a day and average humidity of 64%. The

LEFT: An 18th-century print of Havana harbour, as used on a cigar box.

tobacco is grown on many privately owned smallholdings that sell the product at a fixed price to the Communist government. After the revolution of 1959, most of the cultivated land was seized from the international tobacco companies that were mainly American-owned, and peasants, or *vegueros*, now cultivate the patchwork of smallholdings, ranging in size from 2 hectares (5 acres) to a maximum of 60 hectares (150 acres).

Other areas in Cuba that grow fine tobacco leaf are Semi Vuelta, also in the province of Pinar del Rio, the Partidos area close to Havana, Remedios, in the centre of the island and Oriente at the eastern end.

In the mid-19th century there were nearly 10,000 tobacco plantations in Cuba and over

BELOW: The Partagas factory in downtown Havana.

BELOW: Curing barns, where tobacco leaf is left to dry.

ABOVE: Wrapper leaves are hung to air-dry in a curing barn.

1,000 cigar factories situated in Havana and other large cities. At the beginning of the 20th century there were around 120 factories making over 200 brands of cigar in Cuba, and cigar-makers became the core of the Cuban industrial working class. After the revolution, a large proportion of Cuba's leading tobacco and cigar producers fled to other countries, such as the Canary Islands and the Dominican Republic, where many have flourished.

Many claimed that the quality of Havana cigars fell after the revolution, but the Cubans responded by introducing Montecristo and Cohiba cigars, the world's most sought after premium brands. Today, there are only six factories making handmade cigars in Cuba, but production has increased sharply. Over the last few years, Cuba has exported between 50 and 80 million handmade cigars in 22 brands annually – as against around 30 million just after the revolution.

After the revolution the names of the famous Havana factories were changed: La Corona became Fernando Roig; Partagas, Francisco Perez German; El Rey del Mundo, Heroes del Moncada; and the best-known of all, Romeo Y Julieta, became Briones Montoto. However, the old names are still displayed outside the 19th- and early 20th-century Spanish-style factory buildings. Each factory concentrates on the manufacture of a number of brands of a particular flavour. For example, the Partagas, Gloria Cubana, Bolivar and Ramon Allones brands are all made in the old 1845 Partagas factory, which manufactures nearly 5 million cigars a year.

First-class Cuban brands include the very full-bodied Bolivar, the superb Cohiba, the superior Diplomaticos, the rare and mild Gispert, the superior H. Upmann and Hoyo de Monterrey, to name but a few. Generally, Havana cigars have a medium to full-bodied taste, with coffee, honeyed and earthy tones.

BELOW: Workers in the Partagas factory open bunches of dried wrapper leaf.

BOLIVAR

Bolivar cigars were introduced in 1901 by the Rocha Company of Havana. They are named after Simon Bolivar, one of the great romantic figures of the 19th century, who liberated much of South America from oppressive Spanish rule. Although a large range of Bolivar cigars is hand-made, there are also some machine-made versions available. The flavour of the reasonably priced handmade cigars is full, and they are therefore not recommended for the novice cigar smoker. However, the machine-bunched range is not so strong in flavour and is therefore more suitable for the beginner. There is also a separate range of Bolivar cigars made in the Dominican Republic.

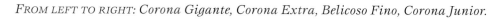

FROM LEFT TO RIGHT: *Corona Gigante, Corona Extra, Belicoso Fino, Corona Junior.*

Name	Size	Ring gauge
Corona Gigante	17.8cm/7in	47
Churchill	17.8cm/7in	47
Palma	17.8cm/7in	33
Immensa	17cm/6¾in	43
Lonsdale	16.8cm/6⅝in	43
Gold Medal Lonsdale	16.2cm/6⅜in	42
Corona Extra	14.3cm/5⅝in	44
Belicoso Fino	14cm/5½in	52
Corona	14cm/5½in	42
Petit Corona	12.7cm/5in	42
Bonita	12.7cm/5in	40
Royal Corona	12.4cm/4⅞in	50
Regente	12.4cm/4⅞in	34
Corona Junior	10.8cm/4¼in	42
Demi-Tasse	9.8cm/3⅞in	30
Cedar-lined Tubes		
Churchill	17.8cm/7in	47
Tubos No. 1	14cm/5½in	42
Tubos No. 2	12.7cm/5in	42
Tubos No. 3	12.4cm/4⅞in	34
Machine-bunched series		
Champion	14cm/5½in	40
Panetela	12.7cm/5in	35
Chicos	10.6cm/4³⁄₁₆in	29

ABOVE: *Bolivar Churchill and cedar-lined tube.*

HABANA

LEFT: *The Bolivar motif from the inside of the cigar box.*

COHIBA

For centuries, historians have speculated on what the name *cohiba* meant to the Cuban Indians of Columbus's time. Some were of the opinion that *cohiba* was a name used for a pipe; others thought it referred to the tobacco plant. It is now known that *cohiba* was the term used for a bunch of tobacco leaves roughly rolled together to form something that we would call a cigar.

The Cohiba brand was created in 1966 as Havana's premier brand, for diplomatic use only. The story goes that Castro's bodyguard had a private supply of cigars which he purchased from a local *tobacquero*. Castro enjoyed the taste of the cigars so much that he secretly installed their maker, Eduardo Ribera, in a Havana mansion, so that he could make them exclusively for the government.

Robusto

From 1982 the brand was offered to the general public in three sizes: Lancero, Corona Especiale and Panetela. Three more sizes were added in 1989 – Esplendido, Robusto and Exquisito – to complete La Linea Clasica. Then in 1992 the five sizes of La Linea 1492 were announced: Siglo I, II, III, IV and V. The flavour of La Linea Clasica is medium to full, while La Linea 1492 is medium.

Cohiba tobaccos are grown on only ten selected *vegas*, or plantations, in the Vuelta Abajo region. The pick of the crop is selected from the five best *vegas* producing, in any year, each type of leaf required to make a Cohiba cigar. Availability will always be limited, because nothing is allowed to compromise the Cohiba

ABOVE LEFT: *Siglo II*
ABOVE RIGHT: *Exquisito*

brand's supremely high standards. Filler tobaccos used for Cohiba cigars benefit from an additional fermentation in cedar casks. This lasts for up to two years and further enhances the delicacy of the tobacco's flavour. It brings to Cohiba a taste that stands out, even in the illustrious company of Havana's other *Grandes Marques.*

Only the most skilful cigar rollers in Cuba are allowed to make Cohibas and, to ensure a perfect result every time, each Cohiba roller specializes in making just one size. Every stage of production is strictly monitored by a highly qualified quality-control team, which is the custodian of the legendary Cohiba aroma and flavour. All finished cigars are checked for weight, girth and shape. Samples of each roller's work are smoked every week, to ensure they satisfy the company's demanding standards for flavour, draw and combustion.

LEFT: *Each box of Cohibas is carefully colour-matched.*

RIGHT: *Panetela*

NAME	SIZE	RING GAUGE
LA LINEA CLASICA SERIES		
Lancero	19cm/7½in	38
Esplendido	17.8cm/7in	47
Coronas Especiale	15.2cm/6in	38
Exquisito	12.7cm/5in	36
Robusto	12.4cm/4⅞in	50
Panetela	11.4cm/4½in	26
LA LINEA 1492 SERIES		
Siglo V	16.8cm/6⅝in	43
Siglo IV	14.3cm/5⅝in	46
Siglo III	15.5cm/6⅛in	42
Siglo II	12.7cm/5in	42
Siglo I	10.2cm/4in	40

FROM LEFT TO RIGHT: Lancero, Esplendido, Siglo V, Siglo III, Siglo IV.

Extra care is taken to colour-match the cigars in each box of Cohibas. They are graded into no less than 80 shades, with a predominance of light- to mid-*claro* tones, and some of a golden *colorado* hue. A range of Cohiba cigars, made in the Dominican Republic by General Cigar, should not be confused with these superb Havanas.

CUABA

The Habanos Corporation introduced the Cuaba brand of mild to medium, handmade cigars to the market in 1996. This new brand recreates a style that was popular at the end of the 19th century and its *figuardos vitolas* were designed especially for smokers who appreciate antiquity. The name Cuaba is very old. The word, like Cohiba, came from the Taino Indians, the original inhabitants of Cuba. It described a special kind of bush, still grown on the island, that burns so well that it was used to light the cigars or *cohibas* during religious ceremonies. Its use was first chronicled in the early 16th century: *"Quemar como una Cuaba"* ("To burn like a Cuaba"). This Cuban expression passed down over the centuries and is still in use today, particularly among farmers.

ABOVE: The Cuaba brand was launched in 1996, but the style of cigar harks back to the 19th century.

RIGHT: Exclusivo
FAR RIGHT: Tradicionale

NAME	SIZE	RING GAUGE
Exclusivo	14.3cm/5⅝in	46
Generoso	13.3cm/5¼in	42
Tradicionale	12cm/4¾in	42
Divino	10.2cm/4in	43

During the 19th century, Havanas with shapely bodies tapering to a point, in a form known as *figuardos*, were the height of fashion. Smaller sizes, too, were popular for enjoyment during intervals at the opera. Gradually, in the 20th century, the fashion changed to *parejos* or straight-sided cigars, and by the late 1930s *figuardos* had virtually disappeared. The new Cuaba *figuardos* are made in the Romeo Y Julieta factory, which has a long tradition of making this style of cigar.

BELOW: *Generoso*
BOTTOM: *Divino*

DIPLOMATICOS

This range of superior, medium- to full-bodied, handmade cigars was introduced to the French market in the 1960s, as a lower-priced alternative to Montecristo. The range comes in seven sizes and availability is somewhat limited, but the cigars represent good value for money. The design of the Diplomaticos cigar band, a horse-drawn carriage, has been widely copied by the Dominican manufacturer of Licenciados cigars.

NAME	SIZE	RING GAUGE
No. 1	16.5cm/6½in	42
No. 2	15.5cm/6⅛in	52
No. 3	14cm/5½in	42
No. 4	12.7cm/5in	42
No. 5	10.2cm/4in	40
No. 6	19cm/7½in	38
No. 7	15.2cm/6in	38

EL REY DEL MUNDO

El Rey del Mundo, or "King of the World", was launched so successfully in 1882 by the Antonio Allones factory, that it was soon renamed the El Rey del Mundo Cigar Company. These popular, handmade cigars are now produced in the Romeo Y Julieta factory in Havana, and the flavour of this large range is light to medium.

NAME	SIZE	RING GAUGE
Grande de Espana	19.2cm/7^9/$_{16}$in	38
Taino	17.8cm/7in	47
Elegante	17.5cm/6^7/$_8$in	28
Lonsdale	16.2cm/6^3/$_8$in	42
Gran Corona	14cm/5^1/$_2$in	46
Corona de Luxe	14cm/5^1/$_2$in	42
Choix Supreme	12.7cm/5in	48
Petit Corona	12.7cm/5in	42
Très Petit Corona	11.4cm/4^1/$_2$in	40
Lunch Club	10.2cm/4in	42
Demi-Tasse	9.8cm/3^7/$_8$in	30

LEFT: *Choix Supreme*   RIGHT: *Lonsdale*

FLOR DE CANO

The company of J. Cano was founded in 1884 by Tomas and José Cano. Against the trend for larger companies, they remained small and independent until the Cuban revolution. Their name lives on in this small selection of light- to medium-flavoured, handmade cigars.

NAME	SIZE	RING GAUGE
Diademas	17.8cm/7in	47
Selectos	14.6cm/5^3/$_4$in	41
Gran Corona	14.3cm/5^5/$_8$in	46
Corona	12.7cm/5in	42
Predilecto Tubulare	12.7cm/5in	42
Short Churchill	12.4cm/4^7/$_8$in	50

HOYO DE MONTERREY

Sr. José Gener grew tobacco in the Vuelta Abajo village of San Juan y Martinez prior to the founding of his brand of cigars in 1865. Hoyo de Monterrey is one of the oldest brands still in existence. The handmade cigars are medium in flavour, while the Le Hoyo series was introduced in the 1970s, in response to the demand for a richer tasting cabinet range.

BELOW: Des Dieux from the Le Hoyo series.

FROM LEFT TO RIGHT: Double Corona, des Dieux, du Dauphin, du Roi, du Prince, Churchill.

ABOVE FROM LEFT TO RIGHT: Du Gourmet, Epicure No.1, Epicure No. 2, du Deputé, du Maire.

	SIZE	RING GAUGE
Double Corona	19.4cm/7⅝in	49
Churchill	17.8cm/7in	47
Jeanne D'Arc	14.3cm/5⅝in	35
Corona	14cm/5½in	42
Short Corona	12.7cm/5in	42
Margarita	12cm/4¾in	26
LE HOYO SERIES (SLIDE-LID BOX)		
Le Hoyo du Gourmet	16.8cm/6⅝in	33
Le Hoyo des Dieux	15.2cm/6in	42
Le Hoyo du Dauphin	15.2cm/6in	38
Le Hoyo du Roi	14cm/5½in	42
Le Hoyo du Prince	12.7cm/5in	40
Le Hoyo du Deputé	10.8cm/4¼in	38
Le Hoyo du Maire	9.8cm/3⅞in	30
CABINET SELECTION (SLIDE-LID BOX)		
Epicure No. 1	14.3cm/5⅝in	46
Epicure No. 2	12.4cm/4⅞in	50

H. UPMANN

Originally a banker, Herman Upmann became so fond of the cigars that were sent to him from Cuba, that he moved to Havana in 1844. There he continued banking and set up as a cigar-maker. The cigar company was taken over by Frankau & Company in 1922. A further change of ownership came about in the mid-1930s, when it was purchased by the company of Menendez y Garcia. Each box of H. Upmann cigars still carries the signature of Herman Upmann.

LEFT: *Sir Winston,* BELOW LEFT: *Grand Corona,* BELOW RIGHT: *Petit Upmann*

NAME	SIZE	RING GAUGE
Sir Winston	17.8cm/7in	47
Seleccion Suprema	17.8cm/7in	33
Upmann No. 1	16.5cm/6½in	42
Cinco Bocas	16.5cm/6½in	42
Upmann No. 2	15.5cm/6⅛in	52
Grand Corona	14.6cm/5¾in	40
Amatista	14.6cm/5¾in	40
Culebras	14.6cm/5¾in	39
Super Corona	14cm/5½in	46
Corona	14cm/5½in	42
Royal Corona	14cm/5½in	42
Cristales	13.5cm/5⁵⁄₁₆in	42
Kings	12.9cm/5¹⁄₁₆in	42
Petit Corona	12.7cm/5in	42
No. 4	12.7cm/5in	42
Petit Upmann	11.5cm/4½in	36
CABINET SELECTION		
Magnum (Slide-lid Box)	14cm/5½in	46
Connoisseur No. 1 (Slide-lid Box)	12.7cm/5in	48
CEDAR-LINED TUBES		
Monarch	17.8cm/7in	47
Corona Major	12.7cm/5in	42
Corona Minor	11.7cm/4⅝in	40
Corona Junior	11.4cm/4½in	36

LA GLORIA CUBANA

These superior-quality, medium- to full-flavoured, handmade cigars were reintroduced in the 1970s and are produced in the Partagas factory in Havana. The limited quantity Medaille d'Or range is supplied in beautifully varnished 8-9-8 boxes.

BELOW: The famous Partagas factory manufactures La Gloria Cubana brand.

NAME	SIZE	RING GAUGE
Especial (Semi Boite Nature)	19cm/7½in	38
Especial No. 2 (Semi Boite Nature)	15.2cm/6in	38
Joyita (Semi Boite Nature)	11.4cm/4½in	26
No. 1	16.5cm/6½in	42
No. 2	15.5cm/6⅛in	52
No. 3	14cm/5½in	42
No. 4	12.7cm/5in	42
No. 5	10.2cm/4in	40
CEDAR-LINED TUBES		
Tubo	15.2cm/6in	42
Petit Tubo	12.7cm/5in	42

MONTECRISTO

Montecristo was introduced in 1935 by the H. Upmann factory. The brand appeared originally in only five sizes, described by number. Other sizes like the A and Especiales were added in the early 1970s. The brand is named after the novel by Alexandre Dumas, *The Count of Montecristo*, and the boxes are decorated with an attractive fleur-de-lis and crossed sword. Montecristo's distinctive flavour has made it one of the most popular Havana brands for over 20 years. It is the best-selling Havana cigar, accounting for nearly 50 % of all handmade cigars exported, with a large proportion going to Spain, France, the UK and Switzerland. The flavour is medium to full.

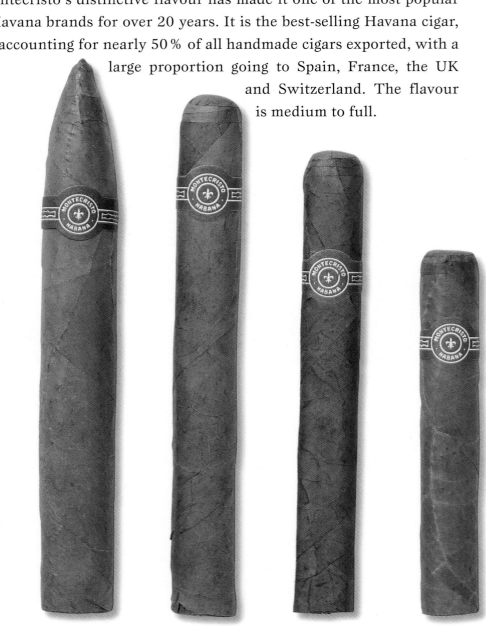

FROM LEFT TO RIGHT: No. 1, No. 2, No. 3, No. 4, No. 5.

FROM LEFT TO RIGHT: Joyita, Especial No. 2, Especial, A.

BELOW: The Montecristo logo is one of the best-known of the Cuban cigars.

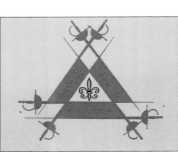

NAME	SIZE	RING GAUGE
A	24cm/9½in	47
Especial	19cm/7½in	38
Especial No. 2	15.2cm/6in	38
Tubos	15.2cm/6in	42
Joyita	11.4cm/4½in	26
No. 1	16.5cm/6½in	42
No. 2	15.2cm/6in	38
No. 3	14cm/5½in	42
No. 4	12.7cm/5in	42
No. 5	10.2cm/4in	42

PARTAGAS

Don Jaime Partagas founded his factory in Havana in 1845. Apart from the period from 1987 to 1990, when the building was being restored, the cigars are still made at this famous factory. This well-known brand of full-flavoured cigars still retains its rich blend of tobaccos. The large range includes many machine-bunched, hand-finished and machine-made examples. A range of Partagas cigars is also handmade in the Dominican Republic and is recognized by the year "1845" printed on the band.

FROM LEFT TO RIGHT: Lusitania, Partagas 8-9-8 (varnished), Partagas 8-9-8 (unvarnished), Corona, Petit Corona.

ABOVE AND RIGHT: *The lining (above)*
and lid (right) of the Partagas box.

LEFT: *Series D No. 4*
RIGHT: *Très Petit Corona*

NAME	SIZE	RING GAUGE
Lusitania	19.4cm/7⅝in	49
Series du Connoisseur No. 1	19.2cm/7⁹⁄₁₆in	36
Churchill De Luxe	17.8cm/7in	47
Palme Grande	17.8cm/7in	33
Partagas de Partagas No. 1	17cm/6¾in	43
Seleccion Privada No. 1	17cm/6¾in	43
Partagas 8-9-8 (Varnished)	17cm/6¾in	43
Lonsdale	16.5cm/6½in	42
Presidente	15.5cm/6⅛in	47
Partagas 8-9-8 (Unvarnished)	15.2cm/6in	42
Corona Grande	15.2cm/6in	42
Culebra	14.4cm/5¹¹⁄₁₆in	39
Corona	14cm/5½in	42
Charlotte	14cm/5½in	35
Petit Corona	12.7cm/5in	42
Series D No. 4 (Semi Boite Nature)	12.4cm/4⅞in	50
Très Petit Corona	11.4cm/4½in	40
Short	10.8cm/4¼in	42
Chico	10.5cm/4⅛in	29

POR LARRANAGA

The oldest Havana cigar still in production, originally introduced in 1834, Por Larranaga cigars are not widely available. Smartly dressed in reddish, oily wrappers, they are of a superior quality, with a medium- to full-bodied flavour. The cigar is mentioned in Rudyard Kipling's poem *The Betrothed*: "There's peace in a Larranaga, there's calm in a Henry Clay". The range of cigars also includes machine-made examples.

BELOW: The old and much-loved Por Larranaga brand features in the writings of Rudyard Kipling. Today, they are available in extremely limited numbers.

NAME	SIZE	RING GAUGE
Lonsdale	16.5cm/6½in	42
Lancero	14.1cm/5⁹⁄₁₆in	42
Corona	14cm/5½in	42
Petit Corona	12.9cm/5¹⁄₁₆in	42
Coronita	12.7cm/5in	38
Nectare No. 4	11.6cm/4⁹⁄₁₆in	40
Small Corona	11.5cm/4½in	40

PUNCH

The Punch brand was founded in 1840 by Don Manuel Lopez of J. Valle & Company and is one of the oldest still being produced today. The name was adopted with the British market in mind, at a time when the humorous magazine, *Punch*, was extremely popular. The familiar character of Mr Punch, smoking a cigar, is still featured on the labels of each box.

The large range of superior-quality cigars, some of which are machine-made, is of medium flavour and reasonably priced. Confusingly, there is also a superior-quality range of Punch cigars made in Honduras.

LEFT: *Petit Corona*

RIGHT: *Double Corona*

BELOW: *The design on the lining paper of the cigar box features the famous Punch character. The brand was aimed specifically at the British market.*

NAME	SIZE	RING GAUGE
Diadema	24cm/9½in	47
Double Corona	19.4cm/7⅝in	49
Churchill	17.8cm/7in	47
Panetela Grande	17.8cm/7in	33
Super Selection No. 1	15.5cm/6⅛in	42
Black Prince	14.3cm/5⅝in	46
Punch Punch	14.3cm/5⅝in	46
Seleccion de Luxe No. 1	14.3cm/5⅝in	46
Corona	14cm/5½in	42
Royal Coronation	14cm/5½in	42
Presidente	12.7cm/5in	42
Petit Corona	12.7cm/5in	42
Margarita	12cm/4¾in	26
Petit Coronation	11.4cm/4½in	40
Coronet	11.4cm/4½in	34
Punchinello	11.4cm/4½in	34
Très Petit Corona	10.8cm/4¼in	42
Petit Punch	10.2cm/4in	40
CEDAR-LINED TUBES		
Churchill	17.8cm/7in	47
Coronation	12.7cm/5in	42
Petit Coronation	11.4cm/4½in	40
Coronet	11.4cm/4½in	34

ABOVE FROM LEFT TO RIGHT: Churchill, Punch Punch, Corona.

ABOVE: The Punch cigar band.

RIGHT: Another lining paper design pays homage to the brand's founder, Manuel Lopez.

MANUEL LOPEZ

QUINTERO

Augustin Quintero opened a small factory in the 1920s in Cienfuegos, near the Remedios tobacco-growing region in the centre of Cuba. The success of his cigars enabled him to found, with his eldest brother, the Quintero y Hermano company in Havana in 1940. These light-flavoured cigars, with their delicate blend of Vuelta Abajo tobaccos, are ideal for a first introduction to Havanas. Some of the Quintero range of cigars are machine-made and appear in the same sizes as the hand-made range.

RIGHT AND BELOW: Although now manufactured in Havana, Quintero packaging makes reference to the company's origins in Cienfuegos.

NAME	SIZE	RING GAUGE
Churchill	16.5cm/6½in	42
Corona	14cm/5½in	42
Nacionale	14cm/5½in	40
Panetela	12.7cm/5in	36
Londres Extra	12.7cm/5in	40
Purito	10.8cm/4¼in	29
CEDAR-LINED TUBES (HAND-FINISHED)		
Tubulare	12.7cm/5in	36

RIGHT: Panetela

RAFAEL GONZALEZ

This range of superior-quality, mild-flavoured, handmade cigars was introduced in 1928 for the British market and is now manufactured in the Romeo Y Julieta factory in Havana. The cedarwood box is printed with the following instructions: "These cigars have been manufactured from a secret blend of pure Vuelta Abajo tobaccos, selected by the Marquez Rafael Gonzalez, Grandee of Spain. For more than twenty years this brand has existed. In order that the Connoisseur may fully appreciate the perfect fragrance, they should be smoked either within one month of the date of shipment from Havana, or should be carefully matured for about one year."

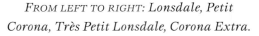

FROM LEFT TO RIGHT: Lonsdale, Petit Corona, Très Petit Lonsdale, Corona Extra.

Name	Size	Ring gauge
Slenderella	17.5cm/6⁷⁄₈in	28
Lonsdale (Vitola "B")	16.2cm/6³⁄₈in	42
Corona Extra	14.3cm/5⁵⁄₈in	46
Petit Corona	12.8cm/5¹⁄₁₆in	42
Petit Lonsdale	12.7cm/5in	42
Panetela Extra	12.7cm/5in	37
Panetela	11.7cm/4⁵⁄₈in	34
Très Petit Lonsdale (Vitola "H")	11.4cm/4¹⁄₂in	40
Cigaritto	11.4cm/4¹⁄₂in	26
Demi-Tasse	10.2cm/4in	30

RAMON ALLONES

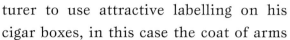

This brand of superior-quality, full-bodied, mainly handmade cigars was founded by a Spaniard, Ramon Allones, in 1837. Allones was the first cigar manufacturer to use attractive labelling on his cigar boxes, in this case the coat of arms

of the Spanish royal family. The cigars, the second oldest in production in Cuba, have been made in the Partagas factory since the 1920s. There are a few machine-made examples also available. A milder range of Ramon Allones is also manufactured in the Dominican Republic.

RIGHT: Ramon Allones packaging makes use of the coat of arms of the Spanish royal family.

FROM LEFT: Petit Corona, Small Club Corona.

NAME	SIZE	RING GAUGE
Gigante	19.4cm/7⅝in	49
Churchill 8-9-8	17cm/6¾in	43
Topper	15.2cm/6in	40
Corona 8-9-8	14.3cm/5⅝in	42
Petit Corona	12.7cm/5in	42
Panetela	12.7cm/5in	35
Allones Specially Selected	12.4cm/4⅞in	50
Ramonita	12.4cm/4⅞in	26
Small Club Corona	11.1cm/4⅜in	42

ROMEO Y JULIETA

Romeo Y Julieta was founded by Alvarez y Garcia in 1875 and purchased by Fernandez Rodriguez in 1903. Rodriguez travelled widely, raced his horse Julieta in Europe and produced many thousands of personalized, one-off cigar bands for the royalty and celebrities of the day. The attractive labelling on the cigar boxes depicts the famous balcony scene from Shakespeare's play *Romeo and Juliet*. The extremely large range of superior-quality cigars, both handmade and machine-made, is medium in flavour. Superior-quality handmade cigars of the same name are also produced in the Dominican Republic.

NAME	SIZE	RING GAUGE
Fabuloso	24cm/9½in	47
Clemenceau	17.8cm/7in	47
Prince of Wales	17.8cm/7in	47
Shakespeare	17.5cm/6⅞in	28
Cedro de Luxe No. 1	16.5cm/6½in	42
Cazadore	16.2cm/6⅜in	44
Corona Grande	15.2cm/6in	42
Belicoso	14cm/5½in	52
Exhibicion No. 3	14cm/5½in	46
Cedro de Luxe No. 2	14cm/5½in	42
Corona	14cm/5½in	42
Petit Corona	12.7cm/5in	42
Romeo No. 2 de Luxe	12.9cm/5¹⁄₁₆in	42
Exhibicion No. 4	12.7cm/5in	48
Cedro de Luxe No. 3	12.7cm/5in	42
Panetela	11.7cm/4⅝in	34
Très Petit Corona	11.4cm/4½in	40
Petit Prince	10.2cm/4in	40
Petit Julieta	10.2cm/4in	30
CEDAR-LINED TUBES		
Churchill	17.8cm/7in	47
No. 1	14cm/5½in	42
No. 2	12.7cm/5in	42
No. 3	11.5cm/4½in	40

ABOVE: *The distinctive red and gold band of Sir Winston Churchill's favourite cigar.*

ABOVE FROM LEFT TO RIGHT: *Prince of Wales, Corona, Cedro de Luxe No. 3, Cedro de Luxe No. 2, Churchill.*

MADE IN HABANA, CUBA

ABOVE: *The balcony scene from* Romeo and Juliet. *Legend has it that this was one of the cigar-makers' favourite plays, which gave name to this popular and well-respected brand.*

BELOW FROM TOP TO BOTTOM: *Petit Prince, Petit Corona, Exhibicion No. 4.*

SAINT LUIS REY

This range of superior-quality, full-bodied, handmade cigars is among the best Havana cigars available. They were introduced in the 1940s and are named after the American film, *The Bridge of Saint Luis Rey*, based on the book by Thornton Wilder. The cigars, made in the Romeo Y Julieta factory in Havana, and packed in smartly designed, white boxes, are only available in limited numbers but, conversely, are reasonably priced. They are not to be confused with the similarly named San Luis Rey cigars from Germany, which are machine-made.

BELOW: *Series A*

BOTTOM: *Petit Corona*

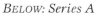

NAME	SIZE	RING GAUGE
Double Corona	19.4cm/7⁵⁄₈in	49
Churchill	17.8cm/7in	47
Lonsdale	16.5cm/6¹⁄₂in	42
Series A	14.3cm/5⁵⁄₈in	46
Corona	14.3cm/5⁵⁄₈in	42
Petit Corona	12.8cm/5¹⁄₁₆in	42
Regios	12.7cm/5in	48

SANCHO PANZA

This small range of superior-quality, mild-flavoured, handmade cigars has only been available in Spain until recently. It is popular with some aficionados as a daytime cigar.

ABOVE: Non Plus

BELOW LEFT AND RIGHT: The box lining depicts the character from Cervantes' famous novel, Don Quixote.

NAME	SIZE	RING GAUGE
Sancho	23.5cm/9¼in	47
Corona Gigante	17.8cm/7in	47
Molino	16.5cm/6½in	42
Dorado	16.5cm/6½in	42
Panetela Largo	16.5cm/6½in	28
Corona	14.3cm/5⅝in	42
Belicoso	14cm/5½in	52
Non Plus	12.7cm/5in	42
Bachillere	11.6cm/4⁹⁄₁₆in	40

TRINIDAD

The rarest brand in the world, the Trinidad is only produced in one size. It is thought by some that Fidel Castro has them made to give as personal gifts to visiting heads of state. This superb-quality, medium- to full-bodied cigar is handmade at the El Laguito factory in Havana. From early 1998 there were plans to make this coveted cigar available to the general public.

NAME	SIZE	RING GAUGE
Trinidad	19cm/7½in	38

RIGHT: Trinidad

VEGAS ROBAINA

Vegas Robaina is a completely new brand of handmade cigars introduced by Habanos S.A. in 1997. Initially only available in Spain, the brand will soon be sold throughout the world. It is named after Don Alejandro Robaina, the head of a renowned Cuban tobacco-growing family. Sr. Robaina has been running their plantations since 1950 and his family's tradition goes back to the mid-19th century. The brand, produced at the H. Upmann factory in Havana, is totally handmade, with filler from the San Luis area of the Vuelta Abajo and the wrapper from the famous Vega Alejandro.

NAME	SIZE	RING GAUGE
Don Alejandro	19.3cm/7⅝in	49
Clasico	16.5cm/6½in	42
Unico	15.5cm/6⅛in	52
Familiar	14cm/5½in	42
Famoso	12.7cm/5in	48

THE DOMINICAN REPUBLIC

The Dominican Republic occupies the eastern two-thirds of the island of Hispaniola in the West Indies. The remaining third is Haiti. Since its discovery by Christopher Columbus in 1492, it has been ruled by a mixture of Spanish, French and Haitians.

The earliest recorded mention of Dominican tobacco concerns Thomas Warner who, in 1622, established tobacco plantations on the north of the island, on behalf of various English companies. Throughout the 18th century tobacco production flourished, with Santiago de los Caballeros the centre of the industry. Spain purchased all the produce of its American colonies, and this policy of monopoly state purchase was also applied to Hispaniola. However, unlike other colonies which were under complete Spanish control, Hispaniola was partially occupied by the French, and the north-west coast was used as a

ABOVE: Tobacco fields in Santiago de los Caballeros.

base by pirates and smugglers. Tobacco producers from the Cibao region preferred to sell their tobacco at market prices to the French and to smugglers. The Cibao producers rarely exported even the minimum quota of tobacco that the Spanish guaranteed to purchase, nor could they declare this "contraband" production, as it was illegal for them to trade with any other country. Dominican tobacco sold in this way was therefore branded as originating from other countries.

The Dominican Republic has had an unstable history. The Spanish ceded the country to France and it was then incorporated into

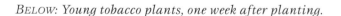

BELOW: Young tobacco plants, one week after planting.

Haiti. Dominica engaged in wars of liberation against the Haitians.

During this period it was independent for nearly 50 years, prior to declaring its own independence in 1844, whereupon the Haitian invasions recommenced. The US army occupied the country from 1914 to 1924, and again in 1965. The Dominican Republic has been ruled by a number of dictators, notably Trujillo (1930–1961), who virtually made tobacco production his own state monopoly.

Tobacco production suffered greatly until after Trujillo's assassination, when, in 1962, the Instituto del Tabaco was formed and immediately set about placing the manufacturing of tobacco on a more professional basis. It selected the best native varieties of tobacco plants and introduced new varieties, the most notable being Piloto Cubano. By selecting appropriate seeds, the Instituto del Tabaco created a technological base that improved the systems of cultivation and production then in place. It then set about stabilizing production.

Throughout the 18th and 19th centuries there was much mass emigration from Dominica to Cuba. The emigrants took with them the tools of their trade, including their farmed tobacco seeds. After the Cuban Revolution of 1959, this was repeated in reverse and Cuban manufacturers introduced their native tobacco seeds into other tobacco-growing countries.

No other country has been able to produce tobacco from Cuban seed, of the same quality

LEFT: Arturo Fuente produces the incredibly rare Fuente Fuente Opus X® in the Dominican Republic.

as the Dominican Republic. It is thought the present Cuban tobacco is the result of partial, unscientific hybridization with Dominican tobacco over the last two centuries. When this tobacco seed was returned to Dominica after 1959, it found growing conditions that were peculiarly suitable. Such is the quality of the seed tobacco that the Dominican Republic exports it to other premium cigar-producing countries, such as Honduras, Jamaica, Mexico, Nicaragua, the USA, and also to Cuba itself.

The Dominican Republic is now the world's biggest producer of premium cigars, which are made entirely by hand. In 1996 it exported 139 million premium cigars to the USA alone, more than the next four top countries put together, and 48% of the total US imports of

BELOW: The Dominican Instituto del Tabaco has improved methods of cultivation, ensuring consistently good-quality, high-yield crops.

ABOVE: Cigar-making at the Rothmans factory. The Dominican Republic is the world's biggest producer of handmade cigars.

premium cigars during 1996. Estimates for the total 1996 Dominican exports are in excess of 250 million premium cigars.

The boom in cigar production in the Dominican Republic started slowly following the USA embargo on Cuban products. At first, there was an increase in demand for tobacco grown in the country, to be exported as leaf to the USA. However, as the production costs for machine-made cigars in the USA rose, and became higher than those in the Dominican Republic for hand-rolled cigar products, more Dominican manufacturers started to introduce their own brands into the world markets.

Some Dominican brands have the same names as the famous Cuban ones. Other first-class Dominican brands include Arturo Fuente, Avo, Casa Blanca, Davidoff, Dunhill, Valdrych, Santa Damiana and Paul Garmirian. Dominican cigars are by far the most popular in the USA and have a mild, sweet taste.

Until recently only filler tobacco was grown in the Dominican Republic. The wrappers and binders were imported from Cameroon, Honduras, Brazil, Mexico, Ecuador and the USA. In recent years wrappers have been successfully grown in the Dominican Republic, notably on the Fuente family's plantation.

ARTURO FUENTE

The Fuente family originally came from Cuba. The first Fuente, Arturo, emigrated to Tampa, Florida, where he set up the Fuente factory. He was succeeded by his son, Carlos Sr. In 1980 a factory was established in the Dominican Republic with seven employees. Now the family employs 1,800 people in six factories in Santiago, and is run by Carlos Fuente Sr and his son, Carlos Fuente Jr. They are the largest producers of handmade cigars in the country, turning out over 24 million cigars every year.

NAME	SIZE	RING GAUGE
Canone	21.6cm/8½in	52
Royal Salute	19cm/7½in	52
Churchill	19cm/7½in	48
Panetela Fina	17.8cm/7in	38
Double Corona	17cm/6¾in	48
Privada No. 1	17cm/6¾in	46
Corona Imperial	16.5cm/6½in	46
Lonsdale	16.5cm/6½in	42
Flor Fina	15.2cm/6in	46
Cuban Corona	13.3cm/5¼in	44
Petit Corona	12.7cm/5in	38
Epicure	11.5cm/4½in	50
Chico	10cm/4in	32
RESERVA SUPERIOR LIMITADA		
No. 3	13.7cm/5½in	40
No. 4	12.7cm/5in	38
Don Carlos	12.7cm/5in	50
OPUS X® SERIES		
Reserva A	23.5cm/9¼in	47
Double Corona	19.3cm/7⅝in	49
Reserva No. 1	16.8cm/6⅝in	44
Reserva No. 2	16cm/6¼in	52
Petit Lancero	16cm/6¼in	38
Fuente Fuente	14.3cm/5⅝in	46
Robusto	13.3cm/5¼in	50

ABOVE LEFT: No. 4
ABOVE RIGHT: Chico

The company uses a blend of four different types of tobacco for its filler and has pioneered the production of wrapper leaves in the Dominican Republic. These have been used on the Fuente Fuente Opus X® series, which was launched in 1995. The production of these cigars is extremely limited, although the company has extended its tobacco plantations next door to the Château de la Fuente farm in El Caribe which, it is hoped, will double the production of this series. Apart from these cigars, the Fuentes produce a standard range and the Hemingway series of large *figuardos*. The cigars are mainly made using rare *colorado* Cameroon wrappers, although some use natural Connecticut Shade wrappers. They are well-made and blended, with a light to medium flavour. The Fuente brand is one of the most popular in the USA.

FROM LEFT TO RIGHT: Double Corona, Panetela Fina, Petit Corona, Reserva Superior Limitada No. 4, Epicure.

ASHTON

These are choice, handmade cigars, named after their creator, William Ashton Taylor, an English pipe-maker. They are now produced by an American-owned company. The cigars, all wrapped in Connecticut leaf with Cuban-seed Dominican binder, are produced in three different styles with distinct flavours: Ashton, Ashton Aged Cabinet Selection and Ashton Aged Maduro. Due to extra ageing, the Cabinet Selection is the mildest. The Aged Maduro, using broad-leaf wrappers, gives a mellow smoke with a sweet flavour.

FROM LEFT TO RIGHT: Magnum, Corona, Panetela, 8-9-8.

NAME	SIZE	RING GAUGE
ASHTON		
Churchill	19cm/7½in	52
Prime Minister	17.5cm/6⅞in	48
8-9-8	16.5cm/6½in	44
Elegante	16.5cm/6½in	35
Double "R"	15.2cm/6in	50
Panetela	15.2cm/6in	36
Corona	14cm/5½in	44
Cordial	12.7cm/5in	30
Magnum	12.7cm/5in	50
ASHTON AGED CABINET SELECTION		
Cabinet No. 1	23cm/9in	52
Cabinet No. 7	16cm/6¼in	52
Cabinet No. 10	19cm/7½in	52
Cabinet No. 8	17.8cm/7in	50
Cabinet No. 2	17.8cm/7in	46
Cabinet No. 3	15.2cm/6in	46
Cabinet No. 6	14cm/5½in	50
ASHTON AGED MADURO		
No. 60	19cm/7½in	52
No. 50	17.8cm/7in	48
No. 30	17cm/6¾in	44
No. 40	15.2cm/6in	50
No. 20	14cm/5½in	44
No. 10	12.7cm/5in	50

AVO

This superior brand of handmade cigars was created in 1986 by the American musician, Avo Uvezian, writer of the hit song "Strangers in the Night". The cigars are all handmade with Connecticut wrapper and Cuban-seed Dominican binders and fillers. The more expensive XO series undergoes a longer ageing process than the standard range.

FROM LEFT TO RIGHT: No. 2, No. 3, No. 4, No. 6, No. 8.

NAME	SIZE	RING GAUGE
No. 3	19cm/7½in	52
Pyramid	17.8cm/7in	54
No. 4	17.8cm/7in	38
No. 5	17cm/6¾in	46
No. 1	17cm/6¾in	42
No. 6	16.5cm/6½in	36
No. 2	15.2cm/6in	50
Belicoso	15.2cm/6in	50
No. 7	15.2cm/6in	44
No. 8	14cm/5½in	40
No. 9	12cm/4¾in	48
Petit Belicoso	11.5cm/4½in	50
XO Series		
Maestoso	17.8cm/7in	48
Preludo	15.2cm/6in	40
Intermezzo	14cm/5½in	50

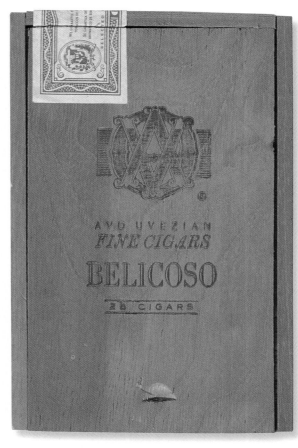

LEFT: *No. 9*   RIGHT AND ABOVE: *Belicoso and box.*

BAUZA

Originally made in pre-Castro Cuba, these superior-quality handmade cigars are made with *colorado* Cameroon wrappers, Mexican binders and Nicaraguan and Dominican fillers. They are medium-bodied, with a rich flavour, and represent good value for money.

ABOVE: All but the Presidente size of Bauza cigars are sold in wooden boxes; Presidentes are sold in bundles of 25.

FROM LEFT TO RIGHT: Casa Grande, Jaguar, Robusto.

NAME	SIZE	RING GAUGE
Presidente	19cm/7½in	50
Medalla D'Oro No. 1	17.5cm/6⅞in	44
Florete	17.5cm/6⅞in	35
Casa Grande	17cm/6¾in	48
Jaguar	16.5cm/6½in	42
Robusto	14cm/5½in	50
Greco	14cm/5½in	42
Petit Corona	12.7cm/5in	38

BUTERA ROYAL

A brand of handmade cigars established in 1993 by pipe-carver, Mike Butera. The unique flavour is created by the use of Java leaf for the binder, adding a slightly spicy taste to the cigars. The wrapper is Connecticut leaf with a Cuban-seed, Dominican filler. The cigars are boxed up with shaved cedarwood, which adds to the flavour.

FROM TOP TO BOTTOM: Capo Grande, Mira Bella, Cedro Fino, Bravo Corto.

NAME	SIZE	RING GAUGE
Capo Grande	19cm/7½in	48
Mira Bella	17cm/6¾in	38
Cedro Fino	16.5cm/6½in	44
Dorado 652	15.2cm/6in	62
Cornetta No. 1	15.2cm/6in	52
Fumo Dulce	14cm/5½in	44
Bravo Corto	11.5cm/4½in	50

CACIQUE

A brand of handmade cigars made in the Dominican Republic by Tropical Tobacco Inc. The filler is a blend of Dominican Havana-seed *ligero* and *seco*, and the cigars have a Havana-seed binder and Connecticut Shade wrapper. This combination gives the brand a full tobacco flavour that is mild to medium in strength. Some sizes are available with a *maduro* wrapper.

NAME	SIZE	RING GAUGE
Inca (No. 8)	19cm/7½in	50
Caribe (No. 7)	17.5cm/6⅞in	46
Siboneye (No. 1)	17cm/6¾in	43
Jaragua (No. 3)	17cm/6¾in	36
Apache	15.2cm/6in	50
Taino (No. 2)	15.2cm/6in	42
Azteca	12cm/4¾in	50

FROM LEFT TO RIGHT: Caribe (No. 7), Apache, Azteca, Siboneye (No.1).

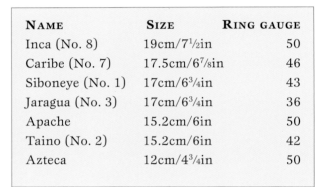

BELOW: Some of the sizes in the Cacique range are available in both maduro *and Connecticut wrappers.*

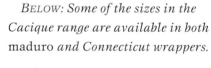

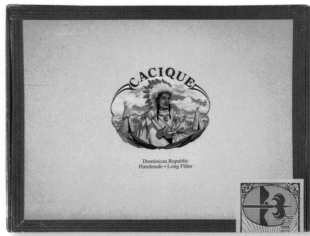

CANARIA D'ORO

The Canaria d'Oro brand is made for the General Cigar Company in Santiago. Production is limited and the cigars are made with wrapper leaves grown in the volcanic soils of the San Andrés Valley region in Central Mexico. The filler tobaccos are a blend of Dominican and Mexican Sumatran tobaccos. The taste of Canaria d'Oro is light, through to medium and full-bodied.

NAME	SIZE	RING GAUGE
Supremo	17.8cm/7in	45
Lonsdale	16.5cm/6½in	42
Fino	15.2cm/6in	31
Immenso	14cm/5½in	49
Corona	14cm/5½in	43
Rothschild	11.5cm/4½in	50
Baby	10.5cm/4⅛in	32

LEFT: *Rothschild Maduro.*

RIGHT: *A box of 50 Rothschild Maduro with the guarantee seal.*

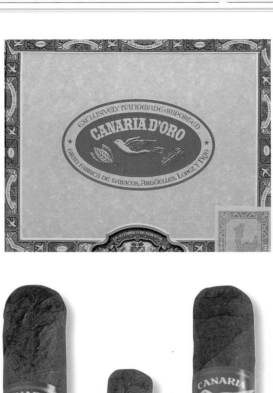

LEFT: *The distinctive "Gold Canary" of the brand's name appears on the box label.*

FROM LEFT TO RIGHT: *Lonsdale, Baby, Corona, Immenso, Fino, Supremo.*

CASA BLANCA

A very mild brand of good-quality, handmade cigars, Casa Blanca (White House) were created for US Presidents. They are all made with Connecticut wrappers, Dominican and Brazilian fillers and Mexican binders. All of the range is available with *claro* or *maduro* wrappers

BELOW: *A box of Casa Blanca cigars, renowned for their size.*

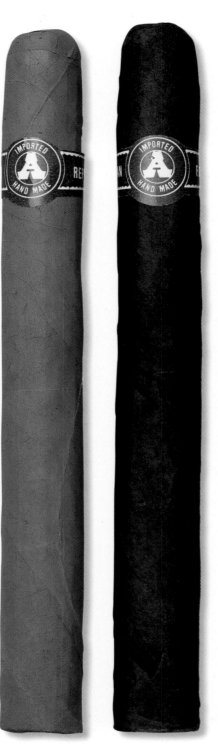

NAME	SIZE	RING GAUGE
Jereboam	25.4cm/10in	66
Presidente	19cm/7½in	50
Magnum	17.8cm/7in	60
Lonsdale	16.5cm/6½in	42
De Luxe	15.2cm/6in	50
Panetela	15.2cm/6in	35
Corona	14cm/5½in	42
Half Jereboam	12.7cm/5in	66
Bonita	10.2cm/4in	36

FROM LEFT TO RIGHT: De Luxe,
De Luxe Maduro, Magnum,
Magnum Maduro, Presidente,
Presidente Maduro.

CHAIRMAN'S RESERVE

This small range of cigars is made at the La Romana factory of Tabacalera de Garcia for the Consolidated Cigar Corporation.

NAME	SIZE	RING GAUGE
Chairman's Reserve	19cm/7½in	38
Churchill	15.2cm/6in	48
Double Corona	17.8cm/7in	50
Robusto	12.7cm/5in	50
Torpedo	15.2cm/6in	50

ABOVE FAR LEFT: Many of these premium cigars are sold in individual cedar boxes.
ABOVE FROM LEFT TO RIGHT: Robusto, Torpedo, Chairman's Reserve.

COHIBA

These cigars were developed by the General Cigar Company in the 1980s to take advantage of the reputation of the legendary Cohiba cigars made in Cuba. The filler is a blend of properly aged tobaccos from the Dominican Republic, with a Jember binder and a Connecticut wrapper. The taste is full-bodied, and the cigars continue to mature in cedar boxes once made.

ABOVE: *The Dominican brand of Cohiba cigars, named after the renowned Cuban brand, is made from a blend of tobaccos grown in the Dominican Republic.*

NAME	SIZE	RING GAUGE
Robusto	12.7cm/5in	49
Churchill	17.8cm/7in	49
Corona Especiale	16.5cm/6½in	42
Lonsdale Grande	15.9cm/6¼in	47
Robusto Fino	12cm/4¾in	47
Corona	13cm/5⅛in	42
Corona Minor	10.2cm/4in	42
Crystal Corona	14cm/5½in	42
Triangulo	15.2cm/6in	54

COHIBA® HAND MADE

ABOVE FROM LEFT TO RIGHT:
Robusto, Corona, Corona Especiale,
Lonsdale Grande, Churchill.

LEFT: Corona Minor

CUESTA-REY

Cuesta-Rey produces three series of superior-quality, handmade cigars – Cabinet Selection, Centennial Collection and No. 95 – in the Dominican Republic. The company was originally founded in Tampa, Florida, in 1884 by Angel La Madrid Cuesta, and the cigars were made from Cuban tobacco imported into the USA. The Centennial Collection, created to celebrate the company's centenary, is made with light Connecticut wrappers and Dominican binders, the Cabinet Selection with Connecticut broad-leaf wrappers, and No. 95 with *colorado* Cameroon wrappers.

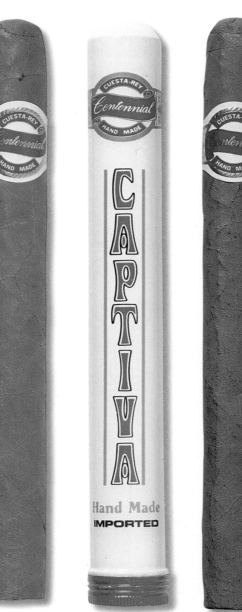

ABOVE: The Cuesta-Rey label illustrates the numerous prizes awarded the brand.

FROM LEFT TO RIGHT: Aristocrat, Captiva and tube, Dominican No.4, Centennial Collection No. 3.

BELOW: A box of Cuesta-Rey Dominican No. 3.

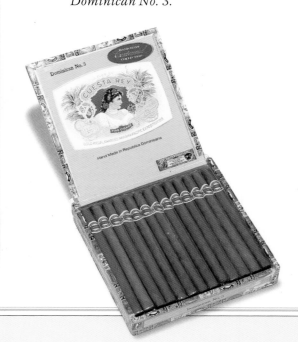

NAME	SIZE	RING GAUGE
CABINET COLLECTION		
No. 1	21.6cm/8½in	52
No. 2	17.8cm/7in	36
No. 95	15.9cm/6¼in	42
No. 898	17.8cm/7in	49
No. 1884	16.5cm/6½in	44
CENTENNIAL COLLECTION		
Individual	21.6cm/8½in	52
Dominican No. 1	21.6cm/8½in	52
Dominican No. 2	18.4cm/7¼in	48
Dominican No. 3	17.8cm/7in	36
Dominican No. 4	16.5cm/6½in	42
Dominican No. 5	14cm/5½in	43
Dominican No. 7	11.4cm/4½in	50
Aristocrat	18.4cm/7¼in	48
Captiva	15.7cm/6³⁄₁₆in	42
Cameo	10.8cm/4¼in	32

FROM LEFT TO RIGHT: No. 5, No. 7 Maduro, No. 7, Cameo, No. 1, No. 1 Maduro.

DAVIDOFF

Zino Davidoff was born in Russia, where his father ran a tobacconist's shop. In 1911 the Davidoff family emigrated to Switzerland and opened a tobacco shop in Geneva. After travelling to South and Central America in the 1920s, Zino ended up in Cuba, where he formed an affectionate relationship with the country that lasted until 1990. After World War II he

FROM LEFT TO RIGHT:
Thousand series: 5000, 4000,
3000, 2000, 1000.

NAME	SIZE	RING GAUGE
Ambassadrice	11.5cm/4½in	26
ANNIVERSARIO SERIES		
No. 1	22.1cm/8¹¹⁄₁₆in	48
No. 2	17.8cm/7in	48
GRAND SERIES		
Grand Cru No. 1	15.5cm/6⅛in	42
Grand Cru No. 2	14.3cm/5⅝in	42
Grand Cru No. 3	12.7cm/5in	42
Grand Cru No. 4	11.7cm/4⅝in	40
Grand Cru No. 5	10.2cm/4in	40
SPECIAL SERIES		
Double R	19cm/7½in	50
Special R	12.7cm/5in	50
Special T	15.2cm/6in	52
THOUSAND SERIES		
1000	11.7cm/4⅝in	34
2000	12.7cm/5in	42
3000	17.8cm/7in	33
4000	15.5cm/6⅛in	42
5000	14.3cm/5⅝in	46
Tubos	15.2cm/6in	38

created his Château selection, based on the Cuban Hoyo de Monterrey brand.

In the early 1970s Zino Davidoff formed a partnership with Ernst Schneider, a major Swiss cigar importer, and Cubatabaco, and

ABOVE: *Davidoff Tubos*

together they successfully marketed three series of Havana cigars throughout the world. In 1990, because of a dispute between Cubatabaco and Schneider's Oettinger Imex company, production of the three series of Davidoffs ceased in Cuba and manufacture was moved to the Dominican Republic. The cigars are now made with *claro* Connecticut wrappers and, although they are among the best cigars produced in Dominica, they are different from the Davidoffs previously produced in Cuba – the overall flavour is lighter.

FROM LEFT TO RIGHT: *Grand Series: Grand Cru No. 1, Grand Cru No. 2, Grand Cru No. 3, Grand Cru No. 4, Grand Cru No. 5.*

LEFT AND RIGHT: *Davidoff cigars are among the best produced in the Dominican Republic.*

FROM LEFT TO RIGHT: *Special T, Special R, Ambassadrice, Anniversario No.2, Anniversario No. 1.*

LEFT: *Davidoff accessories include a wide range of premium cigar cutters.*

DON DIEGO

Originally made in the Canary Islands until the 1970s, this excellent range of handmade cigars uses either Connecticut or Cameroon wrappers, with a Dominican filler and binder. The cigars are made in the Tabacalera de Garcia factory in La Romana for Consolidated Cigar Corporation. Certain sizes are available in either American Market Selection (AMS), double *claro* or English Market Selection (EMS), *colorado* wrappers. These cigars are much appreciated among aficionados and beginners alike for the mild taste.

NAME	SIZE	RING GAUGE
Corona Brava	16.5cm/6½in	48
Corona EMS/AMS	14.3cm/5⅝in	42
Grande	15.2cm/6in	50
Grecos EMS	16.5cm/6½in	38
Lonsdale EMS/AMS	16.8cm/6⅝in	42
Petit Corona EMS/AMS	13cm/5⅛in	42
TUBED CIGARS		
Corona Major Tube EMS	12.8cm/5¹/₁₆in	42
Monarch Tube EMS	18.4cm/7¼in	46
Royal Palm Tube EMS	15.5cm/6⅛in	36
SMALL CIGARS		
Babies	12.8cm/5¹/₁₆in	33
Prelude EMS	10.2cm/4in	28

LEFT: *Lonsdale, Grande, Royal Palm*
RIGHT: *Corona*

DON JUAN PLATINUM

These medium- to full-bodied cigars are a handmade speciality of Tropical Tobacco Inc. in the Dominican Republic. The cigars have a creamy and smooth character, and are made with a Connecticut Shade wrapper, Dominican binder and a Nicaraguan and Dominican filler.

Name	Size	Ring gauge
Presidente	21.6cm/8½in	50
Churchill	17.8cm/7in	49
Numero Uno	16.8cm/6⅝in	44
Torpedo	15.2cm/6in	52 or 30
Matador	15.2cm/6in	50
Cetro	15.2cm/6in	43
Linda	14cm/5½in	38
Robusto	12.7cm/5in	50

LEFT TO RIGHT: Matador, Robusto, Churchill.

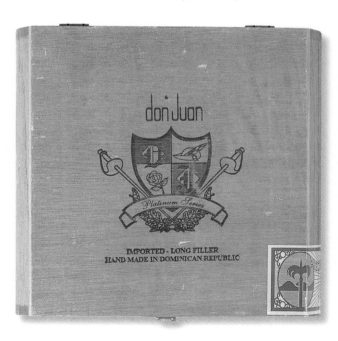

DUNHILL

Dunhill's range of superior-quality, handmade cigars has been manufactured in the Dominican Republic since 1989. For the Dunhill Aged cigars three types of leaf are used: Dominican Piloto, Olor and Brazil. These provide the constituents of the filler and binder; the wrapper is Connecticut Shade. These cigars, distinguished by the blue-and-white Dunhill band, are made from tobacco taken from a single year and are aged in cedar-lined rooms for three months. The properties of the cedarwood allow the aroma and flavour to achieve a subtle balance. A smaller, separate range of Dunhill cigars, recognizable by their black-and-white bands, is made in the Canary Islands.

FROM LEFT TO RIGHT: *Tabaras and tube, Caleta, Bavaro, Romana.*

NAME	SIZE	RING GAUGE
DUNHILL AGED SERIES		
Peravia	17.8cm/7in	50
Cabreras	17.8cm/7in	48
Fantino	17.8cm/7in	28
Diamante	16.8cm/6⅝in	42
Samana	16.5cm/6½in	38
Centena	15.2cm/6in	50
Condado	15.2cm/6in	48
Panetela	15.2cm/6in	30
Tabaras	14.1cm/5⁹⁄₁₆in	42
Valverde	14cm/5½in	42
Altamiras	12.7cm/5in	48
Romana	11.4cm/4½in	50
Bavaro	11.4cm/4½in	28
Caleta	10.2cm/4in	42

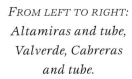

FROM LEFT TO RIGHT:
Altamiras and tube,
Valverde, Cabreras
and tube.

FROM LEFT TO RIGHT: *Peravia, Fantino, Diamante,*
Samana, Condado, Panetela, Senorita.

FONSECA

Fonseca cigars have been made in the Dominican Republic since 1965, although a small range is still made in Cuba. These superior-quality Dominican cigars are handmade using Connecticut wrappers, Mexican binders and Dominican fillers. The cigars are well-made, with a light to medium flavour.

ABOVE: The simple yet distinctive Fonseca logo is stamped on to the lid of every box.

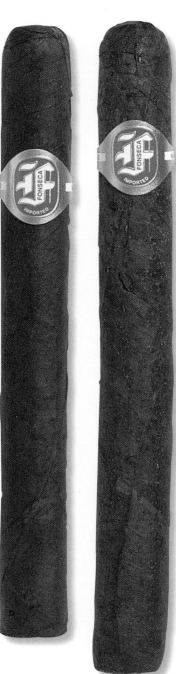

FROM LEFT TO RIGHT: 5-50, 7-9-9, 10-10, Triangular.

NAME	SIZE	RING GAUGE
10-10	17.8cm/7in	50
7-9-9	16.5cm/6½in	46
8-9-9	15.2cm/6in	43
Triangular	14cm/5½in	56
5-50	12.7cm/5in	50
2-2	10.8cm/4¼in	40

THE GRIFFIN'S

This small range of superior-quality and expensive handmade cigars, now solely distributed by Davidoff, has been manufactured in the Dominican Republic for some years. The cigars are made with a Connecticut wrapper and Dominican filler and binder.

NAME	SIZE	RING GAUGE
Prestige	19cm/7½in	50
Privilege	12.7cm/5in	32
No. 100	17.8cm/7in	38
No. 200	17.8cm/7in	43
No. 300	15.9cm/6¼in	43
No. 400	15.2cm/6in	38

FROM LEFT TO RIGHT: Prestige, No. 200, No. 100, No. 300.

BELOW: The mythical griffin appears on all boxes and labels.

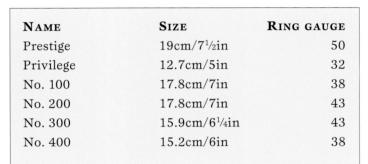

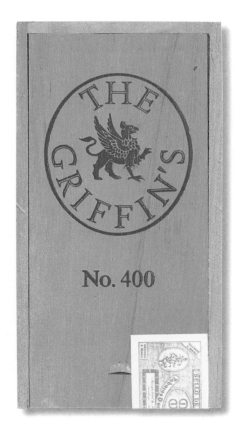

H. UPMANN

H. Upmann, a member of a respected European banking family, opened a cigar factory in Cuba in 1844. After several changes of ownership in 1922 and 1935, the brand is still made in Havana. In more recent times, good-quality Upmann handmade cigars have also been manufactured at the La Romana factory of Tabacalera de Garcia in the Dominican Republic for the Consolidated Cigar Corporation. Until recently the wrappers were Cameroon, but have since been replaced by Indonesian. The binder and filler is Dominican. To differentiate between Dominican and Cuban Upmanns, the labels are marked "H. Upmann 1844" and "H. Upmann Habana" respectively.

FROM LEFT TO RIGHT: Pequenos No. 100, Robusto, Tubo, Naturale Tube, Lonsdale.

NAME	SIZE	RING GAUGE
Churchill	19.4cm/7⅝in	46
Corona Imperiale	17.8cm/7in	46
Corona	14.1cm/5⁹⁄₁₆in	42
Corona Brava	16.5cm/6½in	48
El Prado	17.8cm/7in	36
Lonsdale	16.8cm/6⅝in	42
No. 2000 SBN	17.8cm/7in	42
Petit Corona	12.8cm/5¹⁄₁₆in	42
Pequenos No. 100	11.4cm/4½in	50
Pequenos No. 200	11.4cm/4½in	46
Pequenos No. 300	11.4cm/4½in	42
TUBED CIGARS		
Corona Major Tube	12.8cm/5¹⁄₁₆in	42
Tubo Gold Tube	12.8cm/5¹⁄₁₆in	42
Extra Finos Gold Tube	17cm/6¾in	38
Finos Gold Tube	15.5cm/6⅛in	36
Monarch Tube	17.8cm/7in	46
Naturale Tube	15.5cm/6⅛in	36
Corona Cristal	14.1cm/5⁹⁄₁₆in	42
Panetela Cristal	17cm/6¾in	38
CABINET SELECTION		
Columbo	20.3cm/8in	50
Corsario	14cm/5½in	50
Robusto	12cm/4¾in	50
SMALL CIGARS		
Demi-Tasse	11.4cm/4½in	33
Aperitif	10.2cm/4in	28

BELOW: No. 2000 SBN

HENRY CLAY

These cigars are named after an early 19th-century American congressman and secretary of state, who had business connections with Cuba. They were made in Havana until the 1930s when, for a brief period, the manufacturing moved to New Jersey, USA. The small range of good-quality cigars are now handmade at the Tabacalera de Garcia factory in La Romana for the Consolidated Cigar Corporation, using Connecticut broadleaf wrapper and Dominican filler and binder. The smooth taste is good for a beginner, with different blends created for US and European markets. The boxes still carry the attractive colour picture of the original Henry Clay factory in Havana.

FROM TOP TO BOTTOM: Brevas Finas, Brevas ala Conserva, Brevas.

NAME	SIZE	RING GAUGE
Brevas Finas	16.5cm/6½in	48
Brevas ala Conserva	14.3cm/5⅝in	46
Brevas	14cm/5½in	42

JOSE BENITO

These superior-quality, handmade cigars are made with Cameroon wrappers and have a mild to medium flavour. The company employs less than 70 bunchers and rollers, and the owner, Manuel Queseda, comes from a family which has been in the tobacco business since 1877. The Magnum is one of the largest cigars available in the world and is packed in its own attractive cedar box. The Churchill is appreciated as an after-dinner cigar for its well-balanced flavour, with tones of honey, while the smoothness of the Palma means it is suitable for the beginner, as well as enjoyed by the more experienced smoker.

RIGHT: *José Benito Petite – along with the Palma, it is a suitable cigar for a beginner.*

LEFT: *Rothschild*

NAME	SIZE	RING GAUGE
Magnum	23cm/9in	64
Presidente	19.7cm/7¾in	50
Churchill	17.8cm/7in	50
Corona	17cm/6¾in	43
Panetela	17cm/6¾in	38
Palma	15.2cm/6in	43
Petite	14cm/5½in	38
Havanito	12.7cm/5in	36
Rothschild	11.4cm/4½in	50
Chico	10.2cm/4in	36

FROM LEFT TO RIGHT: *Palma, Panetela, Corona, Churchill.*

JUAN CLEMENTE

Founded by Frenchman Jean Clement in 1982, this company produces fewer than half a million cigars a year, making them difficult to find. The cigars are handmade with Connecticut leaf wrappers, a Dominican binder and a mix of four fillers. This combination of fillers results in a uniquely varied taste. The band is located at the foot of the cigar, holding a protective silver-paper wrapping in place.

FROM LEFT TO RIGHT: *Panetela, Corona, Rothschild, Demi-Corona.*

NAME	SIZE	RING GAUGE
Especiale	19cm/7½in	38
Churchill	17.5cm/6⅞in	46
Panetela	16.5cm/6½in	34
Grand Corona	15.2cm/6in	42
Corona	12.7cm/5in	42
No. 530	12.7cm/5in	30
Rothschild	12.4cm/4⅞in	50
Demi-Corona	10.2cm/4in	40
CLUB SELECTION		
No. 1	15.2cm/6in	50
No. 2	11.4cm/4½in	46
No. 3	17.8cm/7in	44
No. 4	14.6cm/5¾in	42

RIGHT: *Club Selection No. 2,
Club Selection No. 4. The Club Selection series
is aged for 4 years, and offers a more robust
blend than others in the Juan Clemente range.*

BELOW: *When Jean Clement began producing
these cigars in 1982, he used the Spanish
version of his name for the brand.*

KISKEYA

These mild-flavoured cigars are handmade in the Dominican Republic by Tropical Tobacco Inc. They are made with a Dominican filler, Dominican Havana-seed binder and Ecuadorean Connecticut-seed wrapper. They are available with both natural and *maduro* wrapping.

FROM TOP TO BOTTOM: *Churchill, No. 1, Rothschild.*

NAME	SIZE	RING GAUGE
Viajante	21.6cm/8½in	52
Presidente	19cm/7½in	50
Churchill	17.8cm/7in	48
No. 1	17cm/6¾in	44
Palma Fina	17.8cm/7in	36
Cetro	15.9cm/6¼in	44
Toro	15.2cm/6in	50
No. 2	15.2cm/6in	42
No. 4	14cm/5½in	42
Rothschild	12.7cm/5in	50

LICENCIADOS

These superior-quality cigars have been handmade in the Dominican Republic since 1990. The two series of Licenciados cigars are graced by a band bearing the stage-coach emblem, identical to the Cuban brand, Diplomaticos. The cigars are made with Connecticut wrappers and a blend of Dominican filler. These cigars are mild in flavour.

NAME	SIZE	RING GAUGE
Soberano	21.6cm/8½in	52
Presidente	20.3cm/8in	50
Panetela	17.8cm/7in	38
Excelente	17cm/6¾in	43
Toro	15.2cm/6in	50
Licenciados No. 4	14.6cm/5¾in	43
Wavell	12.7cm/5in	50
SUPREME RANGE		
500	20.3cm/8in	50
400	15.2cm/6in	50
300	17cm/6¾in	43
200	14.6cm/5¾in	43

FROM TOP TO BOTTOM: *Supreme 400, Licenciados No. 4, Wavell.*

ABOVE RIGHT: *Supreme 500*

MACANUDO

These superior-quality, handmade cigars, one of the most popular brands in the USA, were first made in Jamaica in 1868. They are now manufactured by the General Cigar Company, both in Jamaica and in the Dominican Republic. The same blends are used in each location and it is extremely difficult to differentiate between the countries of origin. Apart from the Vintage Range, made only in Jamaica, they all use Connecticut Shade wrappers, grown on the company's farms in the Connecticut valley, Mexican binders and a blend of Jamaican, Mexican and Dominican fillers.

Macanudo cigars offer a rich, complex flavour with an exceptionally smooth smoke. They are consistently well-made and deserve their high reputation.

RIGHT: *A box of Macanudo Prince Philip.*

NAME	SIZE	RING GAUGE
Prince of Wales	20.3cm/8in	52
Prince Philip	19cm/7½in	49
Portofino	17.8cm/7in	34
Baron de Rothschild	16.5cm/6½in	42
Duke of Windsor	15.2cm/6in	50
Claybourne	15.2cm/6in	31
Hampton Court	14.6cm/5¾in	43
Crystal	14cm/5½in	50
Hyde Park	14cm/5½in	49
Duke of Devon	14cm/5½in	42
Petit Corona	12.7cm/5in	38
Ascot	10.6cm/4³⁄₁₆in	32
Caviar	10.2cm/4in	36
Miniature	9.5cm/3¾in	24

FROM LEFT TO RIGHT: *Prince of Wales,*
Baron de Rothschild, Crystal,
Duke of Devon, Duke of Windsor,
Hyde Park, Prince Philip Maduro.

RIGHT: *A box of Prince Philip*
Maduro cigars.

MACANUDO VINTAGE

These are also made by the General Cigar Company. Occasionally a tobacco harvest yields a small quantity of exceptional filler, binder and wrapper leaves. These few leaves are given the appellation of "Vintage" and are used to roll Macanudo Vintage Cabinet Selection cigars. The years 1979, 1984, 1988 and 1993 all produced tobacco of this quality, and in the summer of 1996 the first issue of the 1993 vintage crop was released. Macanudo Vintage cigars are identified by the imprint of the year of the vintage on the box and band of each cigar. All cigars have two bands, the red Macanudo Vintage band and a separate, second band showing the vintage from which it came. The leaves are aged more gradually than other tobaccos, and the cigars are incomparably smooth to smoke. The price of these cigars reflects the scarcity and age of the vintage.

FROM LEFT TO RIGHT: Vintage I, Vintage II, Vintage III, Vintage IV.

BELOW: Macanudo Vintage Cabinet selection IV, packaged in a slide-lid cedar box.

NAME	SIZE	RING GAUGE
No. I	19cm/7¹⁄₂in	49
No. II	16.7cm/6⁹⁄₁₆in	43
No. III	14.1cm/5⁹⁄₁₆in	43
No. IV	11.4cm/4¹⁄₂in	47

MONTECRISTO

Not to be confused with the popular cigars of the same name that have been made in Cuba since 1935, this is a more recent brand produced in the Dominican Republic for sale in the USA. The range of seven handmade cigars, apart from the Robusto, all use an American wrapper with a filler and binder from the Dominican Republic. The Robusto uses a wrapper from Cameroon, with a filler and binder from Brazil and the Dominican Republic. The cigars are made at the La Romana factory of Tabacalera de Garcia for the Consolidated Cigar Corporation.

NAME	SIZE	RING GAUGE
Churchill	17.8cm/7in	48
No. 1	16.5cm/6½in	44
Double Corona	15.9cm/6¼in	50
No. 2 (Torpedo)	15.2cm/6in	50
No. 3	14cm/5½in	44
Corona Grande	13.3cm/5¼in	46
Robusto	10.8cm/4¼in	50

RIGHT: Robusto

BELOW: Montecristo cigars, named after the famous Cuban brand, are handmade in the Dominican Republic for the Consolidated Cigar Corporation.

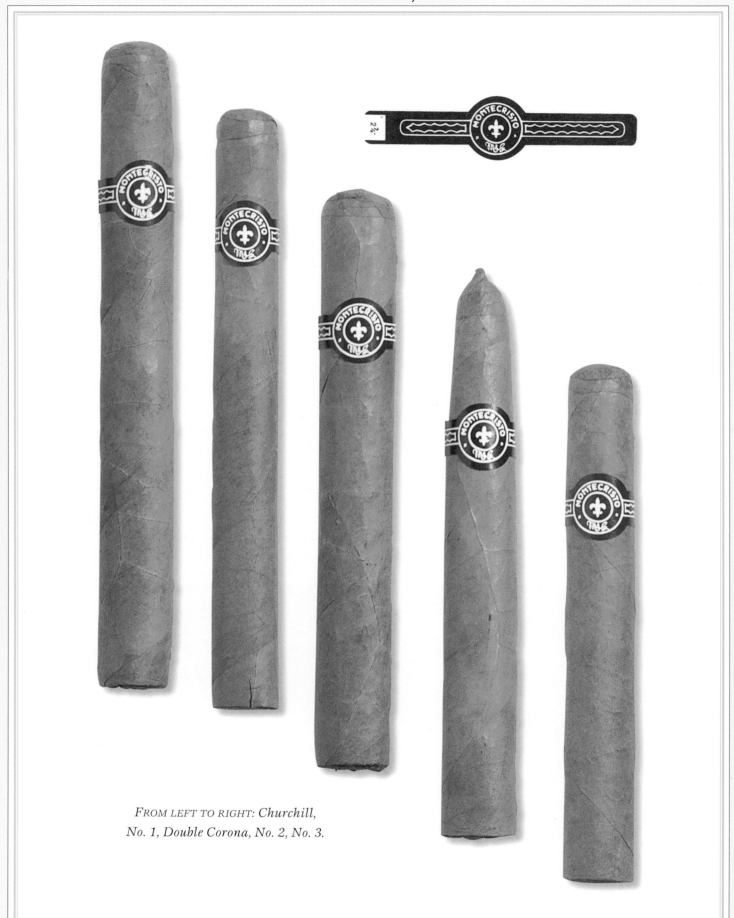

FROM LEFT TO RIGHT: *Churchill,*
No. 1, Double Corona, No. 2, No. 3.

MONTECRUZ

The Menendez family originally made the Montecristo brand in Cuba. Following the Castro Revolution of 1959, the family moved to the Canary Islands, where it started making the Montecruz brand. This operation was moved to the Dominican Republic in the 1970s and these superior-quality cigars are now handmade by the Consolidated Cigar Corporation. The cigars, produced in the Dominican Republic since 1977, are known as "Montecruz Sun Grown". They use Cameroon wrappers with a filler combining Brazilian and Dominican tobaccos and a Dominican binder. A milder range of cigars, using Connecticut wrappers – the Natural Claro Line – is also produced by the company for Dunhill.

NAME	SIZE	RING GAUGE
Individuale	20.3cm/8in	50
Colossus	16.5cm/6½in	50
Tubulare	15cm/1⅛in	36
Cedar-Aged	12.7cm/5in	42
Junior	12.4cm/4⅞in	33
Robusto	11.5cm/4½in	50
Chico	9.8cm/3⅞in	28
Montecruz F	18.4cm/7¼in	47
Montecruz D	17.8cm/7in	36
Montecruz A	16.8cm/6⅝in	43
No. 200	18.4cm/7¼in	46
No. 201	15.5cm/6⅛in	38
No. 205	17.8cm/7in	42
No. 210	16.5cm/6½in	42
No. 220	14cm/5½in	42
No. 240	12cm/4¾in	44
No. 250	16.5cm/6½in	38
No. 255	17.8cm/7in	36
No. 265	14cm/5½in	38
No. 270	12cm/4¾in	35
No. 280	17.8cm/7in	33
No. 281	15.2cm/6in	33
No. 282	12.7cm/5in	42

FROM LEFT TO RIGHT: *Tubulare, Colossus, Junior, Robusto, Cedar-Aged, No.200.*

MONTESINO

This small range of good quality, mild- to medium-flavoured, handmade cigars is manufactured by Arturo Fuente using Cuban-seed Dominican wrappers.

Name	Size	Ring gauge
Napoleon Grande	17.8cm/7in	46
No. 1	17.5cm/6⅞in	43
No. 3	17cm/6¾in	36
Fuma	17cm/6¾in	44
Gran Corona	17cm/6¾in	48
No. 2	15.9cm/6¼in	44
Cesar No. 2	15.9cm/6¼in	44
Diplomatico	14cm/5½in	42

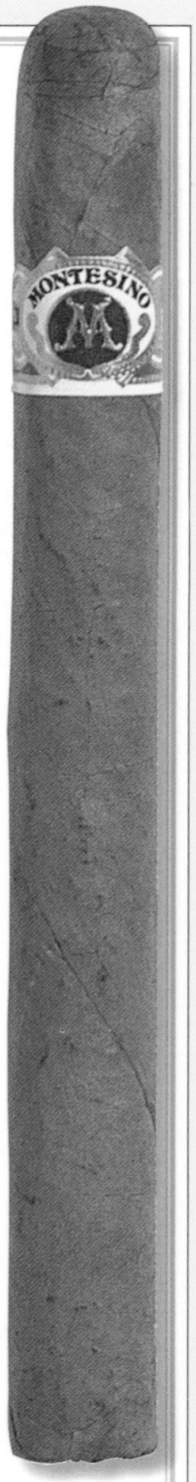

FROM LEFT TO RIGHT:
Diplomatico, Cesar No. 2, No. 2,
Gran Corona, No. 1

NAT SHERMAN

The Nat Sherman tobacco empire, founded in the 1930s and centred on its world-famous shop on Fifth Avenue, in Manhattan, produces an extensive range of handmade Dominican cigars. The distinctive bands carry the emblem of the famous Nat Sherman clock, which dominates the shop front. Different-coloured bands are used to identify each selection, which are all named after the Sherman family or an aspect of New York life.

FROM LEFT TO RIGHT: Metropolitan Selection: Metropolitan, Nautical, University, Explorer, Angler.

The Exchange, Gotham and V.I.P. selections all use Connecticut wrappers, the Landmark selection uses Cameroon wrappers, and the Manhattan and City Desk selections use Mexican wrappers. The Exchange selection has a mild, smooth flavour; the Landmark selection a more full-bodied taste, while the Gotham selection provides a mild smoke. The vast range of Nat Sherman cigars truly provides something for everyone, due to its different combinations of filler, wrapper and binder.

FROM LEFT TO RIGHT: *V.I.P. Selection: Morgan, Carnegie, Astor.*
Gotham Selection: No. 65, No. 711, No. 1400, No. 500.

Name	Size	Ring gauge
METROPOLITAN SELECTION		
Metropolitan	17.8cm/7in	47 or 60
Nautical	17.8cm/7in	34 or 48
University	15.2cm/6in	50
Explorer	14cm/5½in	44 or 56
Angler	14cm/5½in	43
V.I.P. SELECTION		
Morgan	17.8cm/7in	42
Zigfeld Fancytail	17cm/6¾in	38
Carnegie	15.2cm/6in	48
Astor	11.4cm/4½in	50
CITY DESK SELECTION		
Tribune	19cm/7½in	50
Dispatch	16.5cm/6½in	46
Telegraph	15.2cm/6in	50
Gazette	15.2cm/6in	42
HOST SELECTION		
Harrington	19cm/7½in	47
Hampton	17.8cm/7in	50
Hunter	17cm/6¾in	43
Hudson	15.2cm/6in	34
Hobart	14cm/5½in	49
Hamilton	14cm/5½in	42

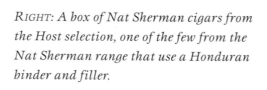

FROM LEFT TO RIGHT: *City Desk Selection: Tribune, Dispatch, Telegraph, Gazette.*

RIGHT: *A box of Nat Sherman cigars from the Host selection, one of the few from the Nat Sherman range that use a Honduran binder and filler.*

Name	Size	Ring gauge
MANHATTAN SELECTION		
Gramercy	17cm/6¾in	43
Chelsea	16.5cm/6½in	38
Tribeca	15.2cm/6in	31
Sutton	14cm/5½in	49
Beekman	13.3cm/5¼in	28
GOTHAM SELECTION		
No. 500	17.8cm/7in	50
No. 1400	15.9cm/6¼in	44
No. 711	15.2cm/6in	50
No. 65	15.2cm/6in	32
EXCHANGE SELECTION		
Butterfield No. 8	16.5cm/6½in	42
Trafalgar No. 4	15.2cm/6in	47
Murray Hill No. 7	15.2cm/6in	38
Academy No. 2	12.7cm/5in	31
LANDMARK SELECTION		
Algonquin	17cm/6¾in	43
Vanderbilt	15.2cm/6in	47
Metropole	15.2cm/6in	34
Hampshire	14cm/5½in	42

ABOVE FROM LEFT TO RIGHT: Landmark Selection: Vanderbilt, Hampshire, Metropole, Algonquin

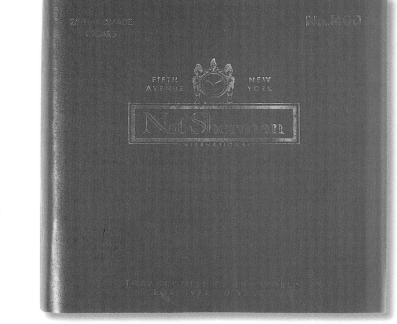

RIGHT: The Gotham selection is packaged in a dark green box, embellished with the famous Nat Sherman clock.

ONYX

This small range of good-quality, handmade, mild cigars was introduced in the early 1990s. They use a Java binder, Dominican and Mexican filler and Mexican *maduro* wrapper.

NAME	SIZE	RING GAUGE
No. 852	20.3cm/8in	52
No. 750	19cm/7½in	50
No. 646	16.8cm/6⅝in	46
No. 642	15.2cm/6in	42
No. 650	15.2cm/6in	50

FROM LEFT TO RIGHT: No. 852, No. 750, No. 642, No. 650.

OSCAR

Introduced in the 1980s, this small range of excellent handmade cigars caters for most tastes. In size, they cover the spectrum from the Oscarito cigarillo to the giant Don Oscar. They all use a mix of Dominican fillers and binders with a Connecticut wrapper.

NAME	SIZE	RING GAUGE
Don Oscar	23cm/9in	46
Supreme	20.3cm/8in	48
No. 700	17.8cm/7in	54
No. 200	17.8cm/7in	44
No. 100	17.8cm/7in	38
No. 300	15.9cm/6¼in	44
No. 400	15.2cm/6in	38
No. 500	14cm/5½in	50
Prince	12.7cm/5in	30
No. 600	11.4cm/4½in	50
Oscarito	10.2cm/4in	20

FROM LEFT TO RIGHT: No.600, No.500, No.400, No.700, No.200, Supreme

PARTAGAS

Partagas was founded in the mid-19th century in Cuba and the original factory in Havana still manufactures an extensive range of these well-known superior cigars. More recently, Partagas cigars have also been handmade in the Dominican Republic by the General Cigar Company. The two countries of origin can be recognized by the bands: the Dominican band is inscribed "Partagas 1845", the Cuban has "Habana".

The Partagas brand in the Dominican Republic is overseen by Ramon Cifuentes, who has spent a lifetime making Partagas cigars, initially in

Cuba, before moving in the 1960s to the Dominican Republic. These cigars are made from Cameroon wrappers, a Mexican binder and a mix of Jamaican, Dominican and Mexican fillers. The finished cigars undergo a unique ageing process for three weeks in a room lined with Spanish cedar. They are then carefully colour-sorted and aged in boxes for another three months. The flavour is full-bodied and rich, with spicy notes and a touch of sweetness.

FROM LEFT TO RIGHT: *No. 10, No. 1, Aristocrat, No. 2, Humitube.*

BELOW: *Partagas Robusto*

Name	Size	Ring gauge
No. 10	19cm/7½in	49
Fabuloso	17.8cm/7in	52
8-9-8	17.5cm/6⅞in	44
Humitube	17cm/6¾in	43
No. 1	17cm/6¾in	43
Maduro	15.9cm/6¼in	48
Almirante	15.9cm/6¼in	47
Aristocrat	15.2cm/6in	50
No. 6	15.2cm/6in	34
Sabroso	14.9cm/5⅞in	44
No. 2	14.6cm/5¾in	43
Naturale	14cm/5½in	50
No. 4	12.7cm/5in	38
Robusto	11.4cm/4½in	49
Purito	10.6cm/4³⁄₁₆in	32
Miniatura	9.5cm/3¾in	24

RIGHT: Robusto, Naturale.

PARTAGAS LIMITED RESERVE

This range is produced by Ramon Cifuentes from the best Partagas Cameroon wrapper leaves, specially aged by a slower maturation process. Partagas claims that these cigars are aged more gradually and for longer periods of time than any other premium cigars. Each cabinet of cigars is accompanied by a certificate that indicates the year and month when the cigars were cedar-aged, packed and released, so the cigar aficionado can be assured of their quality.

LEFT: Epicure

Name	Size	Ring gauge
Royale	17cm/6¾in	43
Regale	15.9cm/6¼in	47
Robusto	14cm/5½in	49
Epicure	12.7cm/5in	38

PAUL GARMIRIAN

Founded in 1991 by the cigar aficionado, Paul Garmirian, the company makes some of the finest handmade cigars outside Havana. These superior-quality cigars are made with reddish-brown *colorado* wrappers and a Dominican binder and filler. They have a sweet and mellow taste and are only available in limited quantities.

LEFT TO RIGHT: *No. 1, Lonsdale, Belicosa, Corona Grande.*

NAME	SIZE	RING GAUGE
Celebration	23cm/9in	50
Double Corona	19.4cm/7⅝in	50
No. 1	19cm/7½in	38
Churchill	17.8cm/7in	48
Belicosa	16.5cm/6½in	52
Corona Grande	16.5cm/6½in	46
Lonsdale	16.5cm/6½in	42
Connoisseur	15.2cm/6in	50
Especial	14.6cm/5¾in	38
Belicoso Fino	14cm/5½in	52
Epicure	14cm/5½in	50
Corona	14cm/5½in	42
Robusto	12.7cm/5in	50
Petit Corona	12.7cm/5in	43
No. 2	12cm/4¾in	48
Petit Bouquet	11.4cm/4½in	38
No. 5	10.2cm/4in	40
Bombone	8.9cm/3½in	43

RIGHT: Petit Bouquet, Double Corona.

LEFT: The Paul Garmirian range is considered by many to provide the best quality cigars outside Havana. Connoisseurs expect to pay accordingly.

PLAYBOY BY DON DIEGO

This small range of English Market Selection (EMS) cigars is made at the La Romana factory of Tabacalera de Garcia for the Consolidated Cigar Corporation. The medium-bodied cigars have a Connecticut wrapper.

FROM LEFT TO RIGHT: Churchill, Gran Corona, Lonsdale, Double Corona, Robusto.

NAME	SIZE	RING GAUGE
Churchill EMS	19.7cm/7¾in	50
Gran Corona EMS	17cm/6¾in	48
Lonsdale EMS	16.5cm/6½in	42
Double Corona EMS	15.2cm/6in	52
Robusto EMS	12.7cm/5in	50

PLEIADES

This range of good-quality, mild cigars is handmade with Connecticut Shade leaf wrapper, Dominican binder and Olor and Piloto Cubano filler. The cigars are made in the Dominican Republic, then sent to France, where they are aged and packed at the company's facility in Strasbourg. The brand is owned by Swisher International. They are then exported throughout Europe and also back across the Atlantic to the USA. The cigars range from mild to rich and full-bodied, depending on the size chosen from the range of 12. The cedar boxes come complete with a built-in humidifier to keep these well-loved cigars in perfect condition.

Pleiades also produces a machine-made cigar, available only in the cigarillo size.

FROM TOP TO BOTTOM: *Orion, Mars.*

NAME	SIZE	RING GAUGE
Aldebran	21.6cm/8½in	50
Saturn	20.3cm/8in	46
Neptune	19cm/7½in	42
Sirius	17.5cm/6⅞in	46
Uranus	17.5cm/6⅞in	34
Orion	14.6cm/5¾in	42
Antares	14cm/5½in	40
Venus	13cm/5⅛in	28
Pluton	12.7cm/5in	50
Perseus	12.7cm/5in	34
Mars	12.7cm/5in	28

FROM LEFT TO RIGHT: *Neptune, Sirius Reserve Privée 1991,*
Sirius, Uranus, Pluton.

POR LARRANAGA

Not to be confused with the Cuban cigar of the same name, these superior-quality, handmade, Dominican cigars are made with Dominican binders, blended Brazilian and Dominican filler and Connecticut Shade leaf wrapper. The Cuban cigars have "Habana" printed on the band, while the Dominican bands are printed "La Romana".

NAME	SIZE	RING GAUGE
Fabuloso	17.8cm/7in	50
Cetro	17.5cm/6^{7}/$_{8}$in	42
Delicado	16.5cm/6^{1}/$_{2}$in	36
Pyramid	15.2cm/6in	50
Nacionale	14.1cm/5^{9}/$_{16}$in	42
Petit Cedro en Cedro	12.7cm/5in	38
Robusto	12.7cm/5in	50

FROM LEFT TO RIGHT: Pyramid, Robusto, Petit Cedro en Cedro, Nacionale, Delicado, Cetro, Fabuloso.

PRIMO DEL REY

This large range of handmade, superior-quality cigars is manufactured at the Tabacalera de Garcia factory in La Romana for the Consolidated Cigar Corporation. The Primo del Rey Club selection is identified by the elaborate gold-and-red band.

FROM TOP TO BOTTOM: Presidente, No.4.

NAME	SIZE	RING GAUGE
Seleccion No. 1	17.3cm/6¹³⁄₁₆in	42
Seleccion No. 3	17.3cm/6¹³⁄₁₆in	36
Presidente	17cm/6³⁄₄in	44
Chavon	16.5cm/6½in	41
Seleccion No. 2	15.9cm/6¼in	42
Reale	15.5cm/6⅛in	36
Cazadore	15.4cm/6¹⁄₁₆in	44
Seleccion No. 4	14cm/5½in	42
Panetela Extra	15.1cm/5¹⁵⁄₁₆in	34
Corto	10.2cm/4in	28
CLUB SELECTION		
Baron	21.6cm/8½in	52
Regal	17.8cm/7in	50
Aristocrat	17cm/6³⁄₄in	48
Noble	15.9cm/6¼in	44
GIFT PACK		
Royal Corona	15.2cm/6in	46
Lonsdale	16.5cm/6½in	42

RAMON ALLONES

Not to be confused with the Cuban brand of the same name, these excellent cigars, handmade by General Cigar in the Dominican Republic, have a mild to medium flavour. They are made with a Cameroon wrapper, Mexican binder and a filler that is a blend of Dominican, Mexican and Jamaican tobaccos. The Crystals are packed in individual glass tubes and the Trumps are packed in a cedar box without band or cellophane. Distribution is limited.

RIGHT: B, Crystal, Redondo.

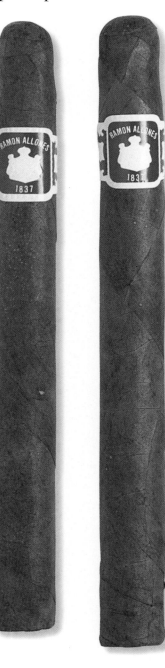

NAME	SIZE	RING GAUGE
Redondo	17.8cm/7in	49
A	17.8cm/7in	45
Trump	17cm/6¾in	43
Crystal	17cm/6¾in	42
B	16.5cm/6½in	42
D	12.7cm/5in	42

ROMEO Y JULIETA

One of the most famous cigar brands in the world, these superior, medium-bodied cigars are now handmade in the Dominican Republic, as well as Cuba. The Dominican cigars are made with a Cameroon wrapper, Connecticut broad-leaf binder and a Dominican and Cuban-seed filler. The excellent Vintage series, packed in a cedar box complete with humidifier, is made with a Connecticut Shade wrapper, an aged Mexican binder, and Dominican and Cuban-seed filler.

NAME	SIZE	RING GAUGE
VINTAGE SERIES		
Monarch	20cm/7⅞in	52
Presidente	17.8cm/7in	43
Delgado	17.8cm/7in	32
Romeo	15.2cm/6in	46
Palma	15.2cm/6in	43
Breva	14cm/5½ in	38
Panetela	13cm/5⅛in	35
Chiquita	10.8cm/4¼in	32

FROM LEFT TO RIGHT:

Palma, Chiquita, Monarch, Presidente, Breva.

ROYAL DOMINICANA

This excellent range of mild- to medium-bodied cigars, handmade for J.R. Tobacco, is made with a Connecticut wrapper, Mexican binder and Dominican filler. The cigars are of excellent quality for their moderate price.

FROM LEFT TO RIGHT: Nacional, Super fino, Corona, No. 1, Churchill.

NAME	SIZE	RING GAUGE
Churchill	18.4cm/7¼in	50
No. 1	17cm/6¾in	43
Corona	15.2cm/6in	46
Super fino	15.2cm/6in	35
Nacional	14cm/5½in	43

ROYAL JAMAICA

This superior range of mild cigars was produced in Jamaica until 1989, when a hurricane flattened the crop and cigar factories. The extensive range was then handmade by the Consolidated Cigar Corporation in the Dominican Republic, using a Cameroon wrapper, a Java binder and a filler made from a blend of Dominican and Jamaican tobacco. From 1996, production began again in Jamaica, in addition to the Dominican manufacture. The *maduro* range is made with Brazilian wrappers, and a secret ingredient added during the fermentation process ensures the uniquely spicy flavour characteristic of Royal Jamaica.

ABOVE: *Gaucho*

NAME	SIZE	RING GAUGE
Goliath	23cm/9in	64
Churchill	20.3cm/8in	51
Giant Corona	19cm/7½in	49
Double Corona	17.8cm/7in	45
Doubloon	17.8cm/7in	30
Navarro	17cm/6¾in	34
Corona Grande	16.5cm/6½in	42
Rapier	16.5cm/6½in	28
Park Lane	15.2cm/6in	47
New York Plaza	15.2cm/6in	40
Royal Corona	15.2cm/6in	30
Buccaneer	14cm/5½in	30
Gaucho	13.3cm/5¼in	33
Robusto	11.4cm/4½in	49
Pirate	11.4cm/4½in	30
MADURO RANGE		
Churchill	20.3cm/8in	51
Corona Grande	16.5cm/6½in	42
Corona	14cm/5½in	40
Buccaneer	14cm/5½in	30

FROM LEFT TO RIGHT: *Corona,
Toro, Park Lane, Corona Grande,
Navarro, Churchill.*

SANTA DAMIANA

In the early years of the 20th century, the Santa Damiana brand, made in Cuba by the American Cigar Company, was popular both in the USA and the UK. In 1986 the Consolidated Cigar Corporation bought American Cigar and re-introduced the brand in 1992. The cigars are made in the modern Tabacalera de Garcia factory in the Duty Free Zone in the island's south-east corner, near Casa de Campo. The factory is one of the most sophisticated in the world with ultra-modern quality control.

Different blends and size names are used for the American Market Selection and English Market Selection brands. The USA sizes, called Seleccion No. 100, No. 700, and so on, contain a

FROM LEFT TO RIGHT: Corona, Petit Corona, Robusto, Torpedo.

lighter blend of tobaccos and have a milder flavour. The English brands have been evolved in conjunction with Hunters & Frankau of London, who have been cigar importers for over 200 years. These brands use the richer-tasting leaves from Dominican fillers and Connecticut Shade wrappers, and are designed for the smoker who prefers something more fully flavoured.

BELOW: *The Santa Damiana brand uses Dominican-grown filler leaves for both its American and English market selection.*

NAME	SIZE	RING GAUGE
AMERICAN MARKET SELECTION (AMS)		
Seleccion No. 800	17.8cm/7in	50
Seleccion No. 100	17cm/6¾in	48
Seleccion No. 700	16.5cm/6½in	42
Seleccion No. 300	14cm/5½in	46
Seleccion No. 500	12.7cm/5in	50
ENGLISH MARKET SELECTION (EMS)		
Churchill	17.8cm/7in	48
Torpedo	15.2cm/6in	52
Tubulare Extra	14cm/5½in	42
Corona	14cm/5½in	42
Robusto	12.7cm/5in	50
Petit Corona	12.7cm/5in	42
Tubulare	12.7cm/5in	42
Panetela	11.4cm/4½in	36

SOSA

This brand of good-quality, reasonably priced, medium- to full-bodied cigars started life in Miami in the 1960s. The cigars have been handmade in the Dominican Republic since the 1970s and are certainly international in their make-up. The wrappers are either Ecuadorean Sumatra seed or Connecticut broad-leaf, the binders are from Honduras and the fillers are Dominican.

FROM LEFT TO RIGHT: *Piramide No. 2,*
Family selection: No. 7, No. 5, No. 4.

NAME	SIZE	RING GAUGE
Magnum	19cm/7½in	52
Piramide	17.8cm/7in	48
Lonsdale	16.5cm/6½in	43
Governor	15.2cm/6in	50
Santa Fe	15.2cm/6in	35
Breva	14cm/5½in	43
Wavell	12cm/4¾in	50
FAMILY SELECTION		
No. 1	17cm/6¾in	43
No. 2	16cm/6¼in	54
No. 3	14.6cm/5¾in	44
No. 4	12.7cm/5in	40
No. 5	12.7cm/5in	50
No. 6	16cm/6¼in	38
No. 7	15cm/6in	50
No. 8	17cm/6¾in	48
No. 9	19.7cm/7¾in	52

BELOW: *The Sosa family selection is made with a Connecticut Shade wrapper and Dominican filler and binder.*

VALDRYCH

Valdrych cigars are made near Santiago in the Dominican Republic, in the heart of the tobacco-growing region of the Cibao valley. The firm started in the 1980s and is a family concern, using tobacco grown on its own farms. Readily available in the USA since the company began, in 1996 Valdrych formed an English company to import Valdrych cigars into the UK.

NAME	SIZE	RING GAUGE
Quisqueya Real	25.4cm/10in	50
Monumento	20.3cm/8in	50
Caballero	19.7cm/7¾in	38
1904	17cm/6¾in	46
Taino	15.2cm/6in	50
Francisco	14cm/5½in	46
Anacaona	14cm/5½in	42
Conde	11.4cm/4½in	52
Carlos	11.4cm/4½in	50
Sublime	10.2cm/4in	42

FROM LEFT TO RIGHT:
Anacaona, Carlos,
Conde, Sublime.

FROM LEFT TO RIGHT: *Francisco, Taino, 1904, Caballero, Monumento, Quisqueya Real.*

LEFT: *The distinctive Valdrych logo on a box of English Market Selection cigars.*

HONDURAS

The Republic of Honduras is a Central American country between Guatemala and Nicaragua. It has a damp, tropical climate with most of the rain falling in the north. The country once had a rich Mayan culture, reaching its peak in the 4th century. The tobacco plant, native to the country, was smoked by the Mayans for religious purposes. Discovered by Christopher Columbus in 1502, the country was soon conquered and settled by the Spanish, and remained under their control until 1821. Since then, frequent revolutions and army coups have marked its history.

In the early 1960s many Cuban émigré cigar producers left Cuba, after Castro's nationalization of the tobacco industry, and set up in Honduras. There has been major investment in the country by local and American tobacco companies, and the tobacco-growing and cigar-manufacturing industry has become increasingly important. Some Honduran brands have the same names as Cuban ones, for example El Rey del Mundo, Hoyo de Monterrey and Punch. Fine Honduran handmade cigars include Don Ramos, Don Tomas, Excalibur, La Invicta and Zino. Honduran cigars tend to have a spicy and robust taste. Although much of the tobacco is grown locally using Dominican, Connecticut and Cuban seed plants, some of the wrappers, binders and fillers are imported from Mexico, Nicaragua and the Dominican Republic.

Below: Tobacco fields in the Rio Copan valley.
The tobacco plant is native to Honduras.

BACCARAT HAVANA

This small range of reasonably priced, good-quality, full-bodied, handmade cigars is made with a Connecticut Shade leaf wrapper, Mexican binder and Honduran Cuban-seed filler. The special ingredients used to seal the cap add a sweet flavour to these cigars.

NAME	SIZE	RING GAUGE
Polo	17.8cm/7in	52
Churchill	17.8cm/7in	50
No. 1	17.8cm/7in	44
Luchadore	15.2cm/6in	43
Petit Corona	14cm/5½in	42
Rothschild	12.7cm/5in	50
Bonita	11.4cm/4½in	30

FROM LEFT TO RIGHT: Bonita, Rothschild, No. 1, Churchill.

BANCES

This brand was originally founded by Francisco Bances in the 1840s in Cuba. Following the Cuban Revolution, the company moved to Tampa, Florida, where machine-bunched cigars are still manufactured. These mild- to medium-flavoured Honduran cigars are handmade with a Honduran filler and a binder from Ecuador. They have spicy overtones and represent great value for money.

NAME	SIZE	RING GAUGE
Corona Immensa	17cm/6¾in	48
Cazadore	15.9cm/6¼in	44
El Prado	15.9cm/6¼in	36
Palma	15.2cm/6in	42
Breva	14cm/5½in	43
Unique	14cm/5½in	38

FROM LEFT TO RIGHT: Palma, Breva, box of Brevas.

BERING

This brand of cigars was founded in 1905 in Tampa, Florida. The machine-bunched cigars are now hand-rolled in Honduras by Swisher International. They are of excellent quality with a medium-bodied flavour.

NAME	SIZE	RING GAUGE
Grande	21.6cm/8½in	52
Immensa	18.1cm/7⅛in	45
Casinos	18.1cm/7⅛in	42
Baron	18.1cm/7⅛in	42
Torpedo	17.5cm/6⅞in	31 or 54
Cazadore	15.9cm/6¼in	45
Gold 1	15.5cm/6⅛in	33
Hispano	15.2cm/6in	50
Plaza	15.2cm/6in	43
Corona Royale	15cm/6in	41
Coronado (natural/candela)	13cm/5⅛in	45
Imperial	13cm/5⅛in	42
Robusto	11.8cm/4¾in	50

FROM LEFT TO RIGHT: Robusto, Imperial and tube, Coronado Candela, Corona Royale and tube.

FROM LEFT TO RIGHT: *Casino, Hispano, Gold 1, Cazadore, Torpedo, Immensa, Baron, Grande.*

LEFT: *Bering's Casino is packaged in a glass tube with a distinctive gold label.*

C.A.O.

This range of good-quality, mild-flavoured, handmade cigars was introduced in 1994. The cigars are made with Connecticut Shade leaf wrappers, local binder tobacco and a filler of Mexican and Nicaraguan tobacco, producing a smooth, sweet taste.

NAME	SIZE	RING GAUGE
Churchill	20.3cm/8in	50
Presidente	19cm/7½in	54
Lonsdale	17.8cm/7in	44
Corona Gorda	15.2cm/6in	50
Corona	15.2cm/6in	42
Small Corona	12.7cm/5in	40
Robusto	11.4cm/4½in	50

FROM LEFT TO RIGHT:
Robusto, Corona Gorda,
Lonsdale, Churchill.

V CENTENNIAL

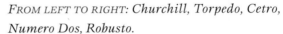

V Centennial are handmade in Honduras by Tropical Tobacco Inc. Selection, and processing of this range began in 1992. It was introduced in November 1993 to commemorate the Fifth Centennial anniversary of the discovery of America and of tobacco by Christopher Columbus. The superior-quality, medium- to full-bodied, handmade cigars are made with Mexican binder, a filler blended from Nicaraguan, Dominican and Honduran tobaccos and a Connecticut Shade leaf wrapper.

FROM LEFT TO RIGHT: Churchill, Torpedo, Cetro, Numero Dos, Robusto.

Name	Size	Ring gauge
Presidente	20.3cm/8in	50
Numero Uno	19cm/7½in	38
Torpedo	17.8cm/7in	36 or 54
Churchill	17.8cm/7in	48
Cetro	15.9cm/6¼in	44
Numero Dos	15.2cm/6in	50
Corona	14cm/5½in	42
Robusto	12.7cm/5in	50

CUBA ALIADOS

This large range of excellent-quality, medium-bodied, handmade cigars is produced in Honduras by Cuban émigré, Rolando Reyes. They are made with Ecuadorean Sumatran-seed wrappers, a Honduran binder, and Dominican and Brazilian filler. At 45.7cm (18in) long, with a ring gauge of 66, the General is the largest cigar currently made in the world.

Name	Size	Ring gauge
General	45.7cm/18in	66
Diadema	19cm/7½in	60
Piramide	19cm/7½in	60
Valentino	17.8cm/7in	48
Palma	17.8cm/7in	36
Corona de Luxe	16.5cm/6½in	45
Lonsdale	16.5cm/6½in	42
Toro	15.2cm/6in	54
No. 4	14cm/5½in	45
Rothschild	12.7cm/5in	51
Petit Cetro	12.7cm/5in	36

DON LINO

This extensive range of good-quality, mild to medium, handmade cigars was introduced in the late 1980s. The standard series is made with Honduran filler and a Connecticut Shade leaf wrapper. The Havana Reserve series is aged for four years and the recently introduced Colorado series has Connecticut broad-leaf wrappers.

NAME	SIZE	RING GAUGE
STANDARD SERIES		
Supremo	21.6cm/8½in	52
Churchill	19cm/7½in	50
Torpedo	17.8cm/7in	48
No. 3	15.2cm/6in	36
Corona	14cm/5½in	50
No. 4	12.7cm/5in	42
Rothschild	11.4cm/4½in	50
Epicure	11.4cm/4½in	32
COLORADO SERIES		
Presidente	19cm/7½in	50
Lonsdale	16.5cm/6½in	44
Robusto	14cm/5½in	50
HAVANA RESERVE SERIES		
Churchill	19cm/7½in	50
Torpedo	17.8cm/7in	48
Tubo	16.5cm/6½in	44
Toro	14cm/5½in	46
Robusto	12.7cm/5in	50
Rothschild	11.4cm/4½in	50
ORO SERIES		
Panetela	17.8cm/7in	36
No. 1	16.5cm/6½in	44
Toro	14cm/5½in	46

ABOVE FROM LEFT TO RIGHT: Oro No. 1, Colorado, Havana Reserve Tubo.
LEFT: A box of Don Lino with selected cigars from the Colorado, Havana Reserve and Oro series.

DON RAMOS

A range of superior-quality, medium- to full-bodied, handmade cigars that is as popular in the UK as it is in the USA. They are made with Honduran fillers, binders and wrappers, creating a spicy flavour.

NAME	SIZE	RING GAUGE
Gigante	17cm/6¾in	47
No. 11	17cm/6¾in	47
No. 13	14.3cm/5⅝in	46
No. 14	14cm/5½in	42
Corona	14cm/5½in	42
Petit Corona	12.7cm/5in	42
No. 16	12.7cm/5in	42
No. 19	11.4cm/4½in	50
No. 20	11.4cm/4½in	42
Très Petit Corona	10.2cm/4in	42
No. 17	10.2cm/4in	42
CEDAR-LINED TUBES		
Churchill	17cm/6¾in	47
Tube No. 1	14cm/5½in	42
Tube No. 2	12.7cm/5in	42
Tube No. 3	11.4cm/4½in	42
Epicure	11.4cm/4½in	50

FROM LEFT TO RIGHT: No. 1 and tube, Petit Corona.

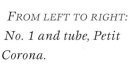

LEFT: The Don Ramos brand is widely available outside the USA.

DON TOMAS

This large range of superior-quality, full- to medium-bodied handmade cigars is available in three different series. The Special Edition series is made with Honduran tobacco grown from Connecticut, Dominican and Cuban seeds. The International series is made with a blend of Cuban-seed tobaccos

NAME	SIZE	RING GAUGE
STANDARD SERIES		
Gigante	21.6cm/8½in	52
Presidente	19cm/7½in	50
Panetela Larga	17.8cm/7in	38
Rothschild	11.4cm/4½in	50
SPECIAL EDITION SERIES		
No. 100	19cm/7½in	50
No. 200	16.5cm/6½in	44
No. 300	12.7cm/5in	50
No. 400	17.8cm/7in	36
No. 500	14cm/5½in	46
INTERNATIONAL SERIES		
No. 1	16.5cm/6½in	44
No. 2	14cm/5½in	50

FROM TOP TO BOTTOM: 400, 500, 300.

EL REY DEL MUNDO

El Rey del Mundo, translated as "King of the World", is an extensive range of superior-quality, full- to medium-bodied, handmade cigars produced in Honduras for J.R. Tobacco. They are not to be confused with the cigars of the same name made in Cuba – their flavour is stronger than this brand. The Honduran examples are made with an Ecuadorean Sumatran-seed wrapper and local filler and binder tobaccos. A lighter-flavoured series is made with a Connecticut Shade leaf wrapper, Dominican filler and locally grown binder tobacco. The current range has been available since 1994, and there are plans to include up to 47 different sizes.

Name	Size	Ring gauge
Coronation	21.6cm/8½in	52
Principale	20.3cm/8in	47
Robusto Supreme	18.4cm/7¼in	54
Corona Immensa	18.4cm/7¼in	47
Double Corona	17.8cm/7in	49
Flor de la Vonda	16.5cm/6½in	52
Choix Supreme	15.5cm/6⅛in	49
Robusto Larga	15.2cm/6in	54
Corona	14.3cm/5⅝in	45
Habana Club	14cm/5½in	42
Robusto	12.7cm/5in	54
Rothschild	12.7cm/5in	50
Lights Series		
Plantation	16.5cm/6½in	30
Elegante	14.3cm/5⅝in	29
Tino	14cm/5½in	38
Reynita	12.7cm/5in	38
Très Petit Corona	12cm/4¾in	43
Petit Lonsdale	11.7cm/4⅝in	43
Café au Lait	11.4cm/4½in	35

FROM LEFT TO RIGHT: Café au Lait, Robusto, Reynita, Tino, Elegante.

FROM LEFT TO RIGHT: Flor del Mundo, Corona Immensa,
Choix Supreme, Del Mundo Tino, Petit Lonsdale.

EXCALIBUR

This range of superior-quality, full- to medium-bodied cigars is sold in the USA as the Hoyo de Monterrey Excalibur brand. Elsewhere they only bear the Excalibur label. Some people consider them to be among the best hand-made cigars available outside Cuba.

NAME	SIZE	RING GAUGE
No. I	18.4cm/7¼in	54
No. II	17cm/6¾in	47
No. III	15.5cm/6⅛in	48
No. IV	14.3cm/5⅝in	46
No. V	15.9cm/6¼in	45
No. VI	14cm/5½in	38
No. VII	12.7cm/5in	43

FROM LEFT TO RIGHT: No. VII, No. VI, No. IV, No. III, No. II, No. I.

FELIPE GREGORIO

This small range of extremely good-quality, medium-flavoured, handmade, *puro* cigars was successfully introduced to the American market in 1990. The cigars are made from 100% Honduran tobacco, producing a smooth flavour.

Name	Size	Ring gauge
Glorioso	19.7cm/7¾in	50
Suntuoso	17.8cm/7in	48
Robusto	12.7cm/5in	52
Belicoso	15.2cm/6in	52
Sereno	14.6cm/5¾in	42
Nino	10.8cm/4¼in	44

FROM LEFT TO RIGHT: Belicoso, Glorioso, Suntuoso, Sereno, Robusto, Nino.

HABANA GOLD

Habana Gold comprises three different ranges of superior-quality, medium- to mild-flavoured, handmade cigars. All are made with a Nicaraguan binder and filler. The Sterling Vintage series has vintage Ecuadorean wrappers, the White Label series has a Nicaraguan wrapper and the Black Label series has an Indonesian wrapper.

NAME	SIZE	RING GAUGE
Presidente	21.6cm/8½in	52
Double Corona	19cm/7½in	46
Churchill	17.8cm/7in	52
No. 2	15.5cm/6⅛in	52
Corona	15.2cm/6in	44
Robusto	12.7cm/5in	50
Torpedo	15.2cm/6in	52
Petit Corona	12.7cm/5in	42

FROM LEFT TO RIGHT: Robusto, Torpedo, Churchill, Petit Corona.

BELOW: Habana Gold cigars are individually wrapped in cellophane within their cedar box.

HOYO DE MONTERREY

Hoyo de Monterrey cigars, made in Honduras, are some of the finest in the world. They differ from the famous Cuban brand of the same name in having a stronger, fuller flavour, particularly in the larger sizes. The range is extensive and of excellent quality. The cigars are full-bodied to medium in flavour, and are made with a filler blended from Honduran, Nicaraguan and Cuban-seed Dominican tobaccos, a Connecticut binder and a wrapper made from Sumatran-seed, Ecuadorean tobacco. The standard range should not be confused with the Excalibur series, which is very different in flavour.

Name	Size	Ring gauge
Presidente	21.6cm/8½in	52
Sultan	18.4cm/7¼in	54
Cuban Largo	18.4cm/7¼in	47
Cetro	17.8cm/7in	43
Double Corona	17cm/6¾in	48
No. 1	16.5cm/6½in	43
Churchill	15.9cm/6¼in	45
Ambassador	16cm/6¼in	44
Governor	15.5cm/6⅛in	50
Delight	15.9cm/6¼in	37
Culebra	15.2cm/6in	35
Corona	13.6cm/5⅜in	46
Café Royal	13.6cm/5⅜in	43
Dreams	14.6cm/5¾in	46
Super Hoyo	14cm/5½in	44
No. 5	13.3cm/5¼in	43
Sabroso	12.7cm/5in	40
Rothschild	11.4cm/4½in	50

FROM LEFT TO RIGHT: Governor, Ambassador. The Governor is characterized by a rich coffee flavour with chocolate overtones, making it perfect as an after-dinner smoke.

J. R. ULTIMATE

J.R. Tobacco of America is the largest mail-order, retail and wholesale purveyor of cigars in the USA. Founded by Jack Rothman and now run by his son, Lew, the company manufactures superior-quality, handmade cigars in Honduras and the Dominican Republic. Two ranges are handmade in the Dominican Republic: Special Jamaicans with Connecticut leaf wrappers, and Special Coronas with Ecuadorean wrapper and binder, and a filler with a mix of Brazilian, Honduran and Dominican tobacco. The Honduran range of superior-quality, full-bodied to medium-flavoured, handmade cigars are made with blended, locally grown Cuban-seed tobaccos for filler and binder, and a Nicaraguan wrapper.

NAME	SIZE	RING GAUGE
ULTIMATE SERIES		
Estelo	21.6cm/8½in	52
Presidente	21.6cm/8½in	52
Super Cetro	20.9cm/8¼in	43
Cetro	17.8cm/7in	42
Slim	17.5cm/6⅞in	36
Palma Extra	17.5cm/6⅞in	38
Double Corona	17cm/6¾in	48
Padron	15.2cm/6in	54
Corona	14.3cm/5⅝in	45
Small Cetro	14cm/5½in	38
Small Corona	11.7cm/4⅝in	43
Rothschild	11.4cm/4½in	50
No. 1	18.4cm/7¼in	54
No. 5	15.5cm/6⅛in	44
No. 10	20.9cm/8¼in	47

LEFT: *Palma Extra, Slim.*

FROM LEFT TO RIGHT: *Cetro, No. 1,
No. 10, Super Cetro, Presidente.*

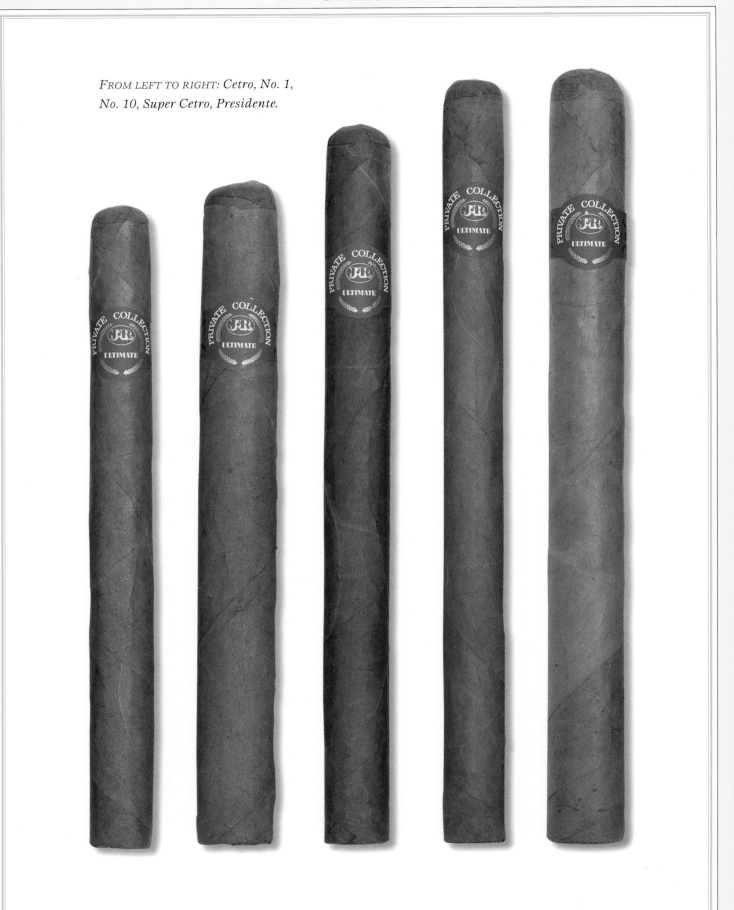

LEMPIRA

This range is made in Honduras by Tropical Tobacco Inc. using a blended filler from Honduras and Nicaragua. The binder is grown in the Dominican Republic from Havana seed and the wrapper is a Connecticut Shade leaf. This is a medium-strength cigar with an abundance of flavour.

Name	Size	Ring gauge
Presidente	19.7cm/7¾in	50
Lancero	19cm/7½in	38
Churchill	17.8cm/7in	48
Lonsdale	16.5cm/6½in	44
Toro	15.2cm/6in	50
Corona	14cm/5½in	42
Robusto	12.7cm/5in	50

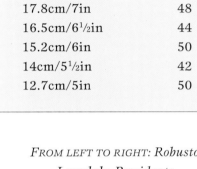

FROM LEFT TO RIGHT: Robusto, Lonsdale, Presidente.

MAYA

A Honduran handmade, long-filler cigar made by Tropical Tobacco Inc., using blended tobaccos from Honduras and Nicaragua. Maya's Havana-seed binder and Connecticut Shade wrapper complete this mild- to medium-strength cigar. The Maya Maduro series is made with the same filler and binder as the Maya Natural, but with a beautiful *maduro* wrapper from Costa Rica.

Name	Size	Ring gauge
Maya Natural Series		
Viajente	21.6cm/8½in	52
Executive	19.7cm/7¾in	50
Torpedo	17.8cm/7in	36 or 54
Elegante	17.8cm/7in	43
Churchill	17.5cm/6⅞in	49
Palma Fina	17.5cm/6⅞in	36
Corona	15.9cm/6¼in	44
Cetro	15.2cm/6in	43
Petit Corona	14cm/5½in	42
Petit	14cm/5½in	34
Robusto	12.7cm/5in	50
Maya Maduro Series		
Executive	19.7cm/7¾in	50
Churchill	17.5cm/6⅞in	49
Corona	15.9cm/6¼in	44
Matador	15.2cm/6in	50
Robusto	12.7cm/5in	50

From left to right: Viajente, Robusto, Corona.

MOCHA SUPREME

This range of good-quality, full- to medium-bodied, handmade cigars is made exclusively with Honduran tobaccos.

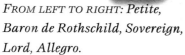

FROM LEFT TO RIGHT: Petite, Baron de Rothschild, Sovereign, Lord, Allegro.

NAME	SIZE	RING GAUGE
Rembrandt	21.6cm/8½in	52
Patron	19cm/7½in	50
Lord	16.5cm/6½in	42
Allegro	16.5cm/6½in	36
Sovereign	14cm/5½in	42
Baron de Rothschild	11.4cm/4½in	52
Petite	11.4cm/4½in	42

PADRON

This range of good-quality, medium- to mild-flavoured, handmade cigars is made with 100% Nicaraguan tobacco in Nicaragua and Honduras. The brand was originally founded in Miami, Florida, in the 1960s.

NAME	SIZE	RING GAUGE
Magnum	23cm/9in	50
Grand Reserve	20.3cm/8in	41
Churchill	17.5cm/6⅞in	46
Panetela	17.5cm/6⅞in	36
Palma	16cm/6⁵⁄₁₆in	42
Londre	14cm/5½in	42
Delicia	12.4cm/4⅞in	46
No. 2000	12.7cm/5in	50
No. 3000	14cm/5½in	52
1964 ANNIVERSARY SERIES		
Diplomatico	17.8cm/7in	50
Monarca	16.5cm/6½in	46
Superior	16.5cm/6½in	42
Corona	15.2cm/6in	42
Exclusivo	14cm/5½in	50

BELOW: No. 2000

210

PARTICULARES

This small range of cigars is made in Honduras by Tropical Tobacco Inc. using Honduran filler. The binder is Havana-seed tobacco grown in Honduras and the wrapper is from Ecuador. This is considered a mild- to medium-strength cigar.

NAME	SIZE	RING GAUGE
Viajante	21.6cm/8½in	52
Presidente	19.7cm/7¾in	50
Supremo	17.8cm/7in	43
Churchill	17.5cm/6⅞in	49
Panetela	17.5cm/6⅞in	35
Royal Corona	15.9cm/6¼in	43
Matador	15.2cm/6in	50
Petit	14.3cm/5⅝in	34
Numero Cuatro	14cm/5½in	42
Rothschild	12.7cm/5in	50

LEFT: Rothschild

RIGHT: Particulares cigars are packaged in wooden boxes with sliding lids.

PETRUS

Launched in 1990, this range of good-quality, mild, handmade cigars has proved to be very popular. The cigars are made with Connecticut-seed wrapper from Ecuador and Honduran binder and filler.

Name	Size	Ring gauge
Lord Byron	20.3cm/8in	38
Double Corona	19.7cm/7¾in	50
Churchill	17.8cm/7in	50
No. 2	15.9cm/6¼in	44
Palma Fina	15.2cm/6in	38
Corona Sublime	14cm/5½in	46
Antonius	12.7cm/5in	52
Gregorius	12.7cm/5in	42
Rothschild	12cm/4¾in	50
Duchess	11.4cm/4½in	30

FROM LEFT TO RIGHT: Corona Sublime, Churchill, Gregorius, Antonius, Rothschild.

PUNCH

Not to be confused with the well-known Cuban cigars of the same name, the Honduran range of Punch handmade cigars is extremely well-made, with a medium to mild flavour. The cigars are made with a filler blended from Dominican, Nicaraguan and Honduran tobaccos, a Connecticut binder and Sumatran-seed Ecuadorean wrapper. The Grand Cru series is aged for at least three years.

RIGHT: Rothschild

NAME	SIZE	RING GAUGE
STANDARD SERIES		
President	21.6cm/8½in	52
Diadema	18.4cm/7¼in	52
After Dinner	18.4cm/7¼in	45
Elegant	18.1cm/7⅛in	36
Casa Grande	17.8cm/7in	46
Largo Elegante	17.8cm/7in	32
Double Corona	16.8cm/6⅝in	48
Bristol	15.9cm/6¼in	50
Amatista	15.9cm/6¼in	44
Punch	15.5cm/6⅛in	43
Café Royal	14.6cm/5¾in	44
No. 75	14cm/5½in	44
Rothschild	11.4cm/4½in	48
GRAND CRU SERIES		
Prince Consort	21.6cm/8½in	52
Diadema	18.4cm/7¼in	54
Tubed Monarca	17cm/6¾in	48
SELECCION DE LUXE SERIES		
Château Lafite	18.4cm/7¼in	54
Corona	15.9cm/6¼in	45
Château Margaux	14.6cm/5¾in	46

FROM LEFT TO RIGHT: *Amatista, Café Royal, Punch, Largo Elegante, No. 75.*

FROM LEFT TO RIGHT: *Diadema,*
Largo Elegante, Casa Grande,
After Dinner, President.

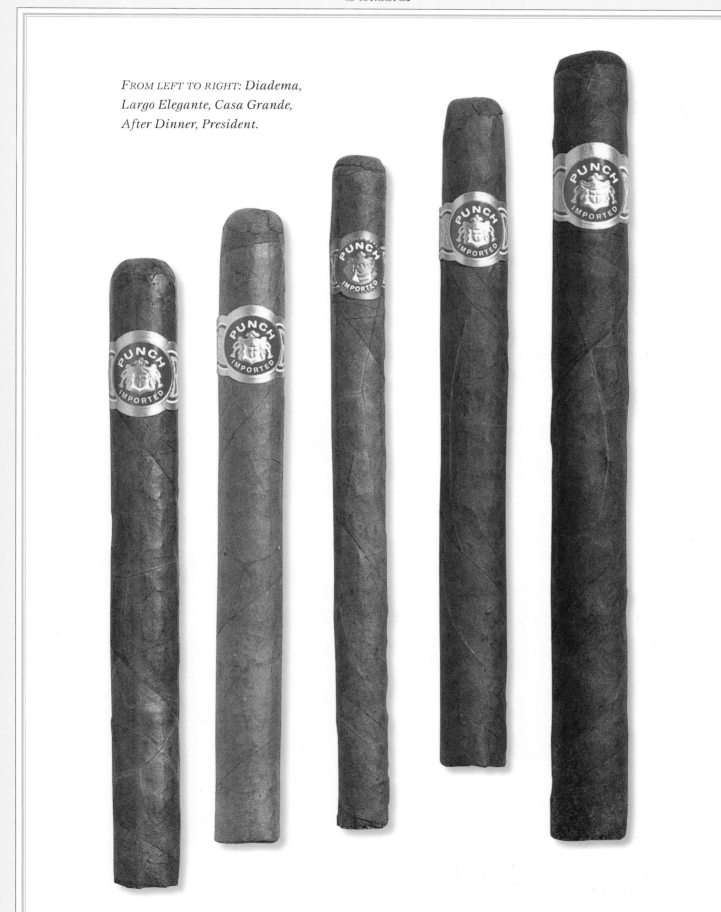

JAMAICA

Jamaica is an independent island nation of the Commonwealth that lies in the West Indies, 90 miles south of Cuba. It was discovered in 1494 by Christopher Columbus and was occupied by the Spanish until the mid-17th century, when British forces successfully invaded and occupied the island. Jamaica remained a British colony until full independence in 1962. It is a mainly mountainous country with a tropical, maritime climate.

Although most people work in agriculture, the principal sources of income are provided by tourism, sugar cane and mining. During the last half of the 19th century many cigar-makers emigrated to Jamaica to escape the oppressive Spanish colonial regime in Cuba. They set up factories in Kingston, and to this day, a few brands of superior-quality, mild, handmade Jamaican cigars are still available for the cigar aficionado. Of these, the best known are the Macanudo brand and the Temple Hall brand.

BELOW: Montego Bay, Jamaica. Although the country produces only a few cigars, they are highly esteemed by the cigar aficionado.

CIFUENTES

The intertwined letters "C" and "R", which form the monogram on the lid of Cifuentes cigars, stand for Ramon Cifuentes, who created them during his tenure as cigar master of the Partagas factory in Havana in 1876. They are now made in Jamaica by General Cigar Company, using three types of Piloto Cubano filler tobaccos grown in the Dominican Republic, and a Jember binder and Connecticut Shade wrapper, grown on the company's own farms. The result is a superior cigar with a deeply mellow flavour.

NAME	SIZE	RING GAUGE
Churchill	18.4cm/7¼in	49
Fancytail	17cm/6¾in	42
Belicoso	15.9cm/6¼in	50
Toro	15.2cm/6in	49
Rothschild	12cm/4¾in	49

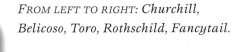

FROM LEFT TO RIGHT: Churchill, Belicoso, Toro, Rothschild, Fancytail.

BELOW: The cigars are packed in boxes originally designed in 1876, with Don Ramon's portrait on the inside lid.

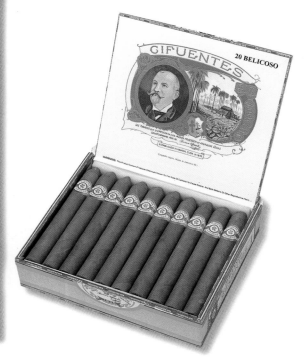

MACANUDO

These superior-quality, handmade cigars, one of the most popular brands in the USA, were first made in Jamaica in 1868. They are now manufactured by the General Cigar Company, both in Jamaica and also in the Dominican Republic. As the same blend of tobaccos is used in both locations, it is extremely difficult to differentiate between their countries of origin. Apart from the Vintage Range, which is made only in Jamaica, Macanudo all use Connecticut wrappers, Mexican binders and Jamaican, Mexican and Dominican fillers. Only Macanudo ages the wrapper leaf a second time, just as it used to be done in Cuba.

LEFT: *Vintage No. II*   RIGHT: *Vintage No. III*

NAME	SIZE	RING GAUGE
Prince of Wales	20.3cm/8in	52
Prince Philip	19cm/7½in	49
Portofino	17.8cm/7in	34
Baron de Rothschild	16.5cm/6½in	42
Duke of Windsor	15.2cm/6in	50
Claybourne	15.2cm/6in	31
Hampton Court	14.6cm/5¾in	43
Crystal	14cm/5½in	50
Hyde Park	14cm/5½in	49
Duke of Devon	14cm/5½in	42
Petit Corona	12.7cm/5in	38
Ascot	10.6cm/4³⁄₁₆in	32
Caviar	10.2cm/4in	36
Miniature	9.5cm/3¾in	24
VINTAGE SERIES		
No. I	19cm/7½in	49
No. II	16.7cm/6⁹⁄₁₆in	43
No. III	14.1cm/5⁹⁄₁₆in	43
No. IV	11.4cm/4½in	47
No. V	14cm/5½in	49

TEMPLE HALL

Temple Hall was a tobacco plantation originally founded by Cuban émigrés in 1876, in the mountains of Jamaica. There the combination of rich soil, dependable rainfall and wind-sheltered fields produced tobacco crops of highly superior quality. The Temple Hall brand was re-introduced by the General Cigar Company in 1992. The excellent quality, mild- to medium-flavoured, handmade cigars are made with a Connecticut Shade wrapper, Mexican binders from the San Andrés Valley region and a filler blended from Dominican, Mexican and Jamaican tobaccos.

NAME	SIZE	RING GAUGE
No. 500	12.7cm/5in	31
No. 550	14cm/5½in	49
No. 625	15.9cm/6¼in	42
No. 675	17cm/6¾in	45
No. 700	17.8cm/7in	49
Belicoso	15.2cm/6in	50
No. 450 Maduro	11.4cm/4½in	49

FROM LEFT TO RIGHT: Belicoso, 675, 500, 700.

RIGHT: Temple Hall, an esteemed Jamaican brand, was re-introduced by General Cigar in recent years.

MEXICO

Long before Christopher Columbus arrived in Central America, the Mayan Indians were smoking rolled-up tobacco leaves. Tobacco growing is now one of the country's main industries and Mexico is a major producer of cigars, including premium types with a mild to heavy flavour. Probably the best known of the Mexican handmade cigars are the Matacan, Santa Clara, Te-Amo and Veracruz brands. The Matacan brand is made by the Consolidated Cigar Corporation in the San Andrés Valley. These cigars are well-made with a medium- to full-bodied flavour. The excellent mild- to medium-flavoured Santa Clara brand was founded in 1830 and is made from 100% Mexican tobacco. Also made by the Consolidated Cigar Company is the Te-Amo range of handmade cigars. Te-Amo are made in the San Andrés Valley in a large range of sizes. Finally, the Veracruz range of superior handmade cigars, also made in the San Andrés Valley, was founded in 1977 by Oscar Franck Terrazas and is sold mainly in the USA. The larger sizes of this brand are packaged in glass tubes and then packed in individual cedar boxes to ensure the cigars' freshness.

BELOW: Mexico produces a small range of highly regarded hand-rolled cigars.

TE-AMO

This range of good-quality, mild to medium, handmade cigars is produced by the Consolidated Cigar Corporation in the tobacco-growing area of the San Andrés valley. Te-Amo (Spanish for "I love you") cigars are very popular in the USA. The range is made from 100% Mexican tobacco.

NAME	SIZE	RING GAUGE
CEO	21.6cm/8½in	52
Gran Piramide	19.7cm/7¾in	54
Presidente	17.8cm/7in	50
Relaxation	16.8cm/6⅝in	44
Toro	15.2cm/6in	50
Satisfaction	15.2cm/6in	46
Meditation	15.2cm/6in	42
Elegante	14.6cm/5¾in	27
Torito	12cm/4¾in	50

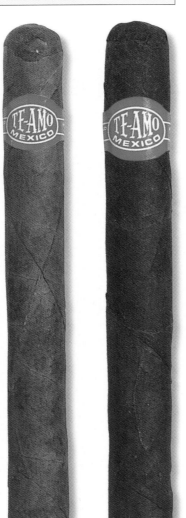

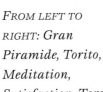

FROM LEFT TO RIGHT: Gran Piramide, Torito, Meditation, Satisfaction, Toro Maduro, Presidente.

VERACRUZ

A range of excellent-quality, mild to medium, handmade cigars produced in the tobacco-growing area of the San Andrés Valley in the state of Veracruz. The brand was founded in the 1970s and is popular in the USA. A great feature of the larger sizes is the packaging. The cigars are sealed in glass tubes with a foam and rubber stopper, and then wrapped in tissue paper before being packed in individual cedar boxes. This ensures that the cigars are fresh, but adds significantly to the cost.

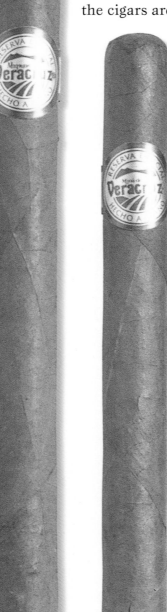

NAME	SIZE	RING GAUGE
Veracruz Magnum	20cm/7⅞in	48
Mina de Veracruz	15.9cm/6¼in	42
Veracruz l'Operetta	12.4cm/4⅞in	34
Poema de Veracruz	15.9cm/6¼in	42
Flor de Veracruz	11.7cm/4⅝in	34

FROM LEFT TO RIGHT: Veracruz Magnum, Poema de Veracruz, Veracruz l'Operetta.

NICARAGUA

Nicaragua is the largest republic in Central America, lying between Honduras and Costa Rica. It has a damp, tropical climate, and about half of the land is under forestation. Most of the sparse population works in agriculture, which includes the growing of tobacco. Discovered in 1502 by Christopher Columbus, the country had a mixture of British and Spanish rulers, and later became part of the Mexican empire. It achieved full independence in 1838. Following a short, self-imposed presidency by an American, William Walker, there followed many years of peace until 1912, when civil war broke out. The country was then occupied by US forces until 1933. From 1937 until the late 1970s, Nicaragua was ruled by the father-and-son presidency of the Somoza family. This was followed by the harsh Sandanista regime and a bloody civil war.

ABOVE: Cutting tobacco on Piedra Azul state farm.

BELOW: Tobacco pickers waiting to be paid. Most of the population of Nicaragua works in agriculture.

Before the civil war, Nicaraguan cigars were considered to be the next best smoke to a Havana. However, the industry suffered terribly in the fighting, with most of the tobacco plantations and barns being destroyed. Since the early 1990s, efforts have been made to re-establish the cigar industry. In particular, the good-quality Joya de Nicaragua brand has shown much improvement, with the maturation of the better local tobaccos used in its manufacture. Other brands of handmade cigars made in Nicaragua include Habanica and Padron. Nicaraguan cigars are full-bodied and aromatic, with a mild to medium taste.

CASA DE NICARAGUA

These handmade cigars are all produced from Nicaraguan-grown Cuban-seed tobacco. The flavour is medium- to heavy-bodied. Many are sold in cedar boxes.

NAME	SIZE	RING GAUGE
Viajante	21.5cm/8³⁄₈in	50
Gigante	20cm/7⁷⁄₈in	54
Presidente	19cm/7¹⁄₂in	52
Churchill	17.8cm/7in	49
Double Corona	17.8cm/7in	44
Panetela Extra	17.8cm/7in	36
Corona	15.2cm/6in	43
Rothschild	12.7cm/5in	50

DON JUAN

A handmade cigar made in Nicaragua by Tropical Tobacco Inc., with a Dominican Havana-seed binder and a Connecticut Shade wrapper. Cigar connoisseurs consider this highly rated cigar medium to full strength. The range is made in eight sizes.

NAME	SIZE	RING GAUGE
Presidente	21.6cm/8¹⁄₂in	50
Churchill	17.8cm/7in	49
Palma Fina	17.5cm/6⁷⁄₈in	36
Numero Uno	16.8cm/6⁵⁄₈in	44
Matador	15.2cm/6in	50
Cetro	15.2cm/6in	43
Linda	14cm/5¹⁄₂in	38
Robusto	12.7cm/5in	50

BELOW: Cetro

HABANICA

This small range of good quality, mild- to medium-flavoured, handmade cigars was introduced in 1995. All of the cigars are made of 100 % Nicaraguan tobacco-grown in the Jalapa Valley.

NAME	SIZE	RING GAUGE
No. 747	17.8cm/7in	47
No. 638	15.2cm/6in	38
No. 646	15.2cm/6in	46
No. 546	13.3cm/5¼in	46
No. 550	12.7cm/5in	50

FROM LEFT TO RIGHT:
550, 638, 646, 747.

JOYA DE NICARAGUA

This range of excellent-quality, medium-bodied, handmade cigars has re-established itself in the marketplace following the disruption caused by the civil war in Nicaragua. They are made with a Connecticut Shade wrapper and locally grown tobacco is used for both filler and binder.

NAME	SIZE	RING GAUGE
Viajante	21.6cm/8½in	52
Churchill	17.5cm/6⅞in	49
No. 1	16.8cm/6⅝in	44
No. 5	17.5cm/6⅞in	35
No. 10	16.5cm/6½in	43
Elegante	16.5cm/6½in	38
No. 6	15.2cm/6in	52
Corona	14.3cm/5⅝in	48
Petit	14cm/5½in	38
Senorita	14cm/5½in	34
Petit Corona	12.7cm/5in	42
Consul	11.4cm/4½in	52
No. 2	11.4cm/4½in	41
Piccolino	10.5cm/4⅛in	30
MADURO DE LUXE		
Presidente	19cm/7½in	54
Toro	15.2cm/6in	50
Robusto	12cm/4¾in	52

FROM LEFT TO RIGHT: Viajante, Petit.

BELOW: No 2

PADRON

This range of good quality, medium- to mild-flavoured, handmade cigars is made with 100% Nicaraguan tobacco in Nicaragua and Honduras. The brand was originally founded in Miami, Florida, in the 1960s. The Anniversary series was introduced in 1994 to commemorate three decades of cigar success.

NAME	SIZE	RING GAUGE
Magnum	23cm/9in	50
Grand Reserve	20.3cm/8in	41
Executive	19cm/7½in	50
Churchill	17.5cm/6⅞in	46
Ambassador	17.5cm/6⅞in	42
Panetela	17.5cm/6⅞in	36
Palma	16cm/6⁵⁄₁₆in	42
No. 3000	14cm/5½in	52
Londres	14cm/5½in	42
Chico	14cm/5½in	36
No. 2000	12.7cm/5in	50
Delicia	12.4cm/4⅞in	46
1964 ANNIVERSARY SERIES		
Diplomatico	17.8cm/7in	50
Monarca	16.5cm/6½in	46
Superior	16.5cm/6½in	42
Corona	15.2cm/6in	42
Exclusivo	14cm/5½in	50

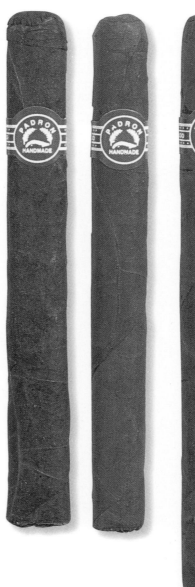

FROM LEFT TO RIGHT: Delicia, Londres, Chico, Palma, Ambassador.

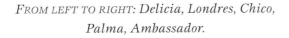

THE PHILIPPINES

In the 17th century a Spanish galleon brought 200 ounces of Cuban tobacco seeds to the Philippines, then a Spanish colony. The seeds were distributed among religious missionaries, who found in the Rio Grande de Cagayan the perfect location for growing tobacco. The tobacco industry flourished in the tropical climate, and 50 years later, a gift of two cases of hand-rolled cigars was presented to the Viceroy of New Spain, as the colony was then known. In 1881, cigar companies in Manila united to establish the largest cigar factory in the Philippines. It was named La Flor de la Isabela, "The Flower of Isabela", after the tobacco cultivated in Isabela, Cagayan. The company continues to manufacture fine hand-made cigars to this day.

BELOW: The tropical climate of the Philippines is ideal for growing tobacco. Brands, such as Double Happiness, are highly regarded by connoisseurs.

DOUBLE HAPPINESS

Double Happiness cigars are produced by the curiously named Splendid Seed Company. They are handmade using prime Filipino binder and filler, which are grown in the lower Cagayan Valley, combined with a double-fermented Brazilian Sumatra wrapper.

Name	Size	Ring gauge
Ecstasy	17.8cm/7in	47
Nirvana	15.2cm/6in	52
Euphoria	16.5cm/6½in	50
Bliss	13cm/5⅛in	48
Rapture	12.7cm/5in	50

FROM TOP TO BOTTOM: Rapture, Bliss, Ecstasy.

LA FLOR DE LA ISABELA

La Flor de la Isabela was formed in 1881, when several cigar factories in Manila united to become the Philippines' largest cigar factory. Since then, the company has manufactured handmade cigars, such as the 1881 and the Tabacalera series, and also produces machine-made cigars.

THE UNITED STATES

Tobacco has been grown in the USA since the scattered tribes from the Mayan civilization travelled to North America. With them they took the tobacco plant, which was subsequently cultivated by native Americans for both medicinal and religious purposes. The first European settlers founded communities in Virginia in 1608, and by 1612 had set up the first tobacco plantations. The locally

BELOW: Some of the best wrappers in the world are grown in the United States.

grown tobacco was smoked in pipes, until the first consignment of Cuban cigars in the 18th century.

In the early 1760s, Israel Putnam returned home to Connecticut from Cuba, with a selection of Havana cigars and large amounts of Cuban tobacco seed. Connecticut had been a tobacco-growing area since the 17th century, and by the early 19th century, cigar factories had opened around Hartford, using leaves from the plants grown from the imported Cuban seed. At the same time Cuban cigars started to be imported into North America in significant quantities. Nowadays, tobacco grown in Connecticut supplies some of the best wrapper leaves in the world, outside Cuba.

The sandy loam of the Connecticut Valley in New England is ideal for growing premium-quality tobacco. The best wrappers are grown under huge tents and are known as Connecticut Shade. They are expensive to produce and can add as much as $1 (£1.60) to the price of a cigar. Drying and maturing the leaves is the same as in Cuba, except that gas burners give additional heat. They are used in the very best cigars, such as Dominican Davidoffs and Jamaican Macanudos.

By the time of the American Civil War (1861–5), cigar smoking had become widespread throughout the country. Following tax reductions in the 1870s, cigars became more readily available and domestic production increased. By the end of the 19th century, as in Europe, the cigar had become a major status

ABOVE: *Miami, Florida, is the site for the manufacture of La Gloria Cubana, one of the best-known brands of handmade cigars in the USA.*

symbol. To combat the growing popularity of cigarettes, machine production was introduced in the 1920s and this reduced the quantity of handmade cigars produced considerably.

During the last quarter of the 19th century, many Cuban cigar-makers fled from Spanish-controlled Cuba to Tampa and Key West in Florida. Nowadays, most cigars made in the USA are machine-made, such as the Cuesta-Rey series, which is still made in Tampa, and is an excellent example of its kind. Cuesta-Rey was one of the greatest cigar houses of Tampa, founded in 1884. However, Cuesta-Rey handmade cigars are all now made in the Dominican

Republic. Other well-known, machine-made brands include Arango, Directors, Garcia y Vega, Havana Blend (using 100% Cuban tobacco) and King Edward. The best-known handmade brand still made in the USA is the superb La Gloria Cubana range, manufactured in Miami using Ecuadorean wrappers, and fillers and binders from Nicaragua, Ecuador or the Dominican Republic.

The overall cigar market in the USA has experienced rapid growth since 1993, reversing a steady decline in the market from 1964. Led by a growth in mass-market, large and premium cigars, the cigar market has increased at a compound annual rate of 9.8% in unit terms between 1993–6, and has increased at almost twice that rate in retail dollar sales.

CONSOLIDATED CIGAR CORPORATION

Consolidated Cigar Corporation is the leading cigar company in the USA, selling nearly 1 billion cigars a year and commanding 25% of the market. Consolidated Cigar Corporation was formed in 1918 by Julius Lichtenstein, President of the American Sumatra Tobacco Company, leaf specialists, which amalgamated six independent cigar manufacturers. One of the six, G. H.

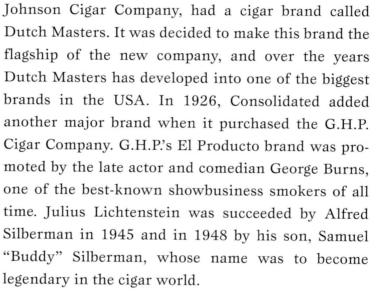

Johnson Cigar Company, had a cigar brand called Dutch Masters. It was decided to make this brand the flagship of the new company, and over the years Dutch Masters has developed into one of the biggest brands in the USA. In 1926, Consolidated added another major brand when it purchased the G.H.P. Cigar Company. G.H.P.'s El Producto brand was promoted by the late actor and comedian George Burns, one of the best-known showbusiness smokers of all time. Julius Lichtenstein was succeeded by Alfred Silberman in 1945 and in 1948 by his son, Samuel "Buddy" Silberman, whose name was to become legendary in the cigar world.

In late 1968 Consolidated was taken over by the Gulf & Western conglomerate. The company then entered the premium, handmade cigar business, through the formation of Cuban Cigar Brands in the Canary Islands with Pepe Garcia, a major Cuban manufacturer, whose factory in Cuba had been nationalized by the Castro regime. It then acquired the Moro Cigar Company and its Primo del Rey trademark.

Gulf & Western sold Consolidated in 1983 to five of its senior managers, and 16 months later the company was sold again to MacAndrews & Forbes, a holding company controlled by Mr Ronald Perelman, chairman of Revlon, Inc. In 1986 it acquired the assets of the American Cigar Company, including Antonio y

LEFT: *Don Diego Lonsdale*, ABOVE RIGHT: *Don Diego Corona*

Cleopatra, La Corona and Roi-Tan cigars. It also purchased the Milton Sherman Tobacco Company and the pipe tobacco brands of Iwan Ries & Company.

In 1988 the President and management purchased Consolidated from MacAndrews & Forbes and immediately made three more acquisitions, Te-Amo cigars, Royal Jamaica Cigars and Century Tobacco Company's pipe tobacco products.

In 1993 the company was re-purchased by MAFCO Holdings Inc., Mr Perelman's personal company, and later in 1996 it went public.

Today the company employs about 5,000 people and has its headquarters in Fort Lauderdale. It operates production facilities in Honduras, Puerto Rico, Pennsylvania, Jamaica, Virginia and the Dominican Republic, as well as Mexico, Brazil and Holland. Leaf supplies are obtained from growers in every part of the globe. The company's trademarks include Antonio y Cleopatra, Don Diego, Dutch Masters, Dutch Treats, El Producto, Flamenco, Henry Clay, H. Upmann, La Corona, Mixture 79, Montecristo, Montecruz, Muriel, Primo del Rey, Roi-Tan, Royal Jamaica, Santa Damiana, Te-Amo and Three Star.

ABOVE FROM LEFT TO RIGHT:
Antonio y Cleopatra Tubos and tube.

FROM LEFT TO RIGHT: Antonio y Cleopatra Grenadier,
Antonio y Cleopatra Paloma Maduro.

OLIVA TOBACCO COMPANY

The Oliva Tobacco Company, with headquarters in Tampa, Florida, supplies tobacco to many of the world's makers of premium cigars, including Punch, La Gloria Cubana and Arturo Fuente. The company now owns 340 hectares (850 acres) of tobacco plantations in Honduras, Ecuador, Nicaragua, the Dominican Republic and the Connecticut River Valley, and employs over 3,000 workers. The Oliva family business started in the Pinar del Rio region of Cuba in the 19th century. In 1924 Angel Oliva, who until his death in 1996 ran the business, emigrated to Florida. In 1934 he launched the Oliva Tobacco Company, and by the 1950s had become one of the major tobacco distributors in the world, handling tobacco from farms in Cuba. Following Castro's Revolution he started producing tobacco in Honduras, and later in other countries in Central and South America. On account of the current cigar boom, demand for the company's tobacco has never been higher, and it is even considering growing tobacco once again in Florida, where costs are obviously higher.

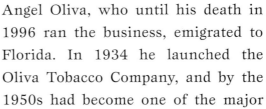

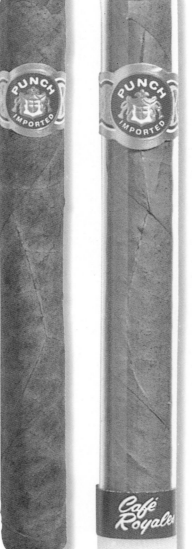

LEFT TO RIGHT: *Punch Amatista, Punch Café Royal, Arturo Fuente Petit Corona, Arturo Fuente Chico.*

TROPICAL TOBACCO INC.

Founded by Pedro Martin, who has had a life-long fascination with cigars, Tropical Tobacco's headquarters are in Miami, Florida. Martin is a legend in the world of cigar blending, and is constantly travelling and fine-tuning the blends that are used to make his cigars. Tropical Tobacco produces nine ranges, all of which are listed under the individual countries. These are V Centennial, V Centennial 500, Don Juan, Don Juan Platinum, Lempira, Maya, Cacique, Kiskeya and Particulares.

*LEFT TO RIGHT: V Centennial Cetros, V Centennial Robusto,
Kiskeya Robusto, Cacique Caribe, Cacique Inca.*

THE ORIGINAL KEY WEST CIGAR FACTORY

In 1965, Eleanor Walsh established the Original Key West Cigar Factory. When it was founded there were a few cigar factories remaining in Key West from the turn of the century. These other factories eventually closed, while the Original Key West Cigar Factory thrived. For the first 25 years the factory employed three people: the owner and two rollers. With each sale, Eleanor distributed a price list and a book of matches and from there the business grew by word of mouth. The company has now expanded through mail order and has many loyal customers. Original customers come back to Key West to visit the factory, and visitors can still see the room where the first cigars were purchased from the Original Key West Cigar Factory.

The company has retained many of the traditional skills associated with handmade cigars, and on any day visitors may view the cigar rollers demonstrating the difficult art of hand-rolling a fine cigar. Three ranges of cigars are made. The long filler cigars are hand-rolled with Cuban-seed Honduran tobacco and an Ecuadorean wrapper, and

LEFT TO RIGHT: Figuardo Torpedo, Jamaican Rhum, House Special.

come in 13 varieties. This combination of tobaccos presents a very mild and smooth cigar. The premium-aged long filler cigars are hand-rolled with an aged Cuban-seed Honduran tobacco, a Cameroon wrapper and come in seven sizes. These cigars are full-bodied and spicy yet mild. The blended cigars are hand-rolled with a blend of various Cuban-seed Honduran tobaccos and come in four sizes. Included in the mild blends are the Rhum cigars which have a slight sweetness and a very aromatic flavour.

LEFT TO RIGHT: Harry Truman, Cayo Hueso Panetela, El Hemingway.

NAME	SIZE	RING GAUGE
LONG FILLER CIGARS		
Key West Caballero	20.3cm/8in	39
El Presidente	20cm/7⅞in	50
Key West Diplomat	17.8cm/7in	48
Cayo Hueso Panetela	17.8cm/7in	36
Carmencita	17.8cm/7in	
Super Fino	17.8cm/7in	
Churchill	17.8cm/7in	50
Elegante	15.2cm/6in	50
Coronado	15.2cm/6in	48
El Hemingway	13.3cm/5¼in	46
House Special	12.7cm/5in	50
Cuban Split	12.7cm/5in	36
Small Conchita	10.2cm/4in	
PREMIUM-AGED LONG FILLER CIGARS		
Harry Truman	17.8cm/7in	48
Conchita Panetela	17.8cm/7in	36
Figuardo Torpedo	15.2cm/6in	56
Gran Corona	15.2cm/6in	48
Mini-Robusto	12.7cm/5in	46
BLENDED CIGARS		
Pirates Alley Panetela	17.8cm/7in	38
Key West Queen	13.3cm/5¼in	50
Jamaican Rhum	12.7cm/5in	44

MASS-MARKET MACHINE-MADE CIGARS

For the aficionado, nothing compares to the quality of a premium hand-rolled cigar, no matter what its country of origin. For the beginner, some mass-market products merit investigation. In the United States, there has long been a tradition of good-quality affordable cigars, while in Europe, consumption of machine-made cigars is enormous.

The EU is not only a major consumer of cigars, but also a major manufacturer of machine-made ones. Among the leading cigar-manufacturing countries are the Netherlands, Spain, Germany, Denmark and Switzerland. During 1996, total cigar and cigarillo consumption in the EU totalled 6.4 billion pieces. The leading consumer was France (1.5 billion) followed by Germany and the United Kingdom.

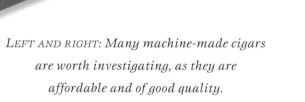

LEFT AND RIGHT: Many machine-made cigars are worth investigating, as they are affordable and of good quality.

THE OFFICIAL E.E.C. DEFINITION OF A CIGAR

Any tobacco product, which can be smoked as it is and which is:

♦ A roll of tobacco with an outer wrapper of natural tobacco; or

♦ A roll of tobacco containing predominantly broken or threshed leaf, with a binder of reconstituted tobacco, and with an outer wrapper which is of reconstituted tobacco, having the normal colour of a cigar and which is fitted spirally; or

♦ A roll of tobacco containing predominantly broken or threshed leaf, with an outer wrapper of reconstituted tobacco, having the normal colour of a cigar; and having a weight, exclusive of any detachable filter or mouthpiece, of not less than 2.3g (0.09oz); and having a circumference over at least one-third of its length, of not less than 34mm ($^{15}/_{16}$in).

BELGIUM

TABACOFINA-VANDER ELST

In 1874 Henri Vander Elst, a farmer's son from Wilsele, opened a specialist tobacco shop. Soon, with the help of his two brothers, François and Alphonse, he widened the scope of his business to encompass the manufacture of different tobacco products: cigars, chewing tobacco, snuff and pipe tobacco. From then on, the company, Vander Elst Frères, expanded and its success enabled factories to be built in and outside Belgium. The Turnhout factory, which was founded in 1945, specializes in cigar production and the Corps Diplomatique brand in particular. This range, with a subtle and mild taste, comes in 12 different sizes. The filler is a blend of tobacco from Java, Brazil, Havana and Sumatra and the wrappers are provided by fine Sumatran leaves.

The Mercator brand of cigars was also created by Vander Elst in 1960 and named after the famous Belgian geographer, Gerhardus Mercator, 1512–94. Mercator is known for his world maps, made for navigation purposes, but his reputation was established by the invention of geographical projections in order to obtain parallel meridians, first used in his map of 1568. The Mercator range of cigars is very wide, from the miniatures to the more traditionally shaped cigars.

FROM LEFT TO RIGHT: Jupiter, Fiesta, Fiesta Natural, Fiesta Mild, Jupiter Mild.

GERMANY

DANNEMANN

Dannemann is a well-known maker of small cigars. The company was founded in Brazil by Geraldo Dannemann in 1873. The cigars are machine-made in Germany, using tobacco imported from Brazil and Sumatra. The brand is now owned by the company Ritmeester Cigars of the Netherlands.

Name	Size	Ring gauge
Vera Cruz Aromatico	16.7cm/6⁹⁄₁₆in	32
Vera Cruz Ligero	16.7cm/6⁹⁄₁₆in	32
Lights Brazil	15.4cm/6¹⁄₁₆in	30
Lights Sumatra	15.4cm/6¹⁄₁₆in	30
Lonja Brazil	13.6cm/5³⁄₈in	20
Lonja Sumatra	13.6cm/5³⁄₈in	20
Espada Brazil	13.2cm/5³⁄₁₆in	44
Espada Sumatra	13.2cm/5³⁄₁₆in	44
Imperial Brazil	10.2cm/4in	20
Imperial Sumatra	10.2cm/4in	20
Moods	7.3cm/2⁷⁄₈in	20
Speciale Brazil	7cm/2³⁄₄in	22
Speciale Sumatra	7cm/2³⁄₄in	22

FROM LEFT TO RIGHT: Lights, Imperial Brazil, Lonja and box.

HOLLAND

The cigar has been a favourite form of smoking in the Netherlands since manufacture started at the end of the 18th century. Due to the dominance of the East Indies by the Dutch, from the beginning of the 17th century to 1945, Dutch cigars have included leaf grown firstly in Java and then Sumatra. Many Dutch cigars of today are composed of about 70% Indonesian leaf, combined with a large proportion of Brazilian and Cuban tobacco leaves. Holland is world-famous for its quality, machine-made brands of the "dry" Dutch type of cigars. All the cigars are now machine-made, some using homogenized tobacco, but many are made with the finest, aged tobaccos from Colombia, the Dominican Republic, Mexico, Java and Cameroon.

The Dutch Cigar Manufacturers Association is formed by six well-known Dutch manufacturers of cigars and cigarillos. Together, these six

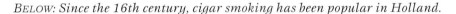

BELOW: Since the 16th century, cigar smoking has been popular in Holland.

manufacturers share over 90% of the Dutch market and almost 100% of the Dutch exports of cigars and cigarillos. The members of the Association and their most important brands are: Agio Sigarenfabrieken N.V. with the brands Agio, Panter, Balmoral and De Huifkar; Cadena Claassen Sigarenfabrieken B.V. with the brands Cadena Claassen, Carl Upmann and Acht Zaligheden; Swedish Match Cigars B.V. with the brands La Paz, Willem II, Karel I, Justus van Maurik and Heeren van Ruydael; Henri Wintermans Sigarenfabrieken B.V. with the brands Henri Wintermans and Café Crème; Chambord with brands Ritmeester B.V., Ritmeester, Oud Kampen, Hajenius and Danneman; and Schimmelpenninck Sigarenfabrieken B.V. which markets the one brand, Schimmelpenninck.

The Dutch cigar industry produced 1.8 billion cigars, señoritas and cigarillos in 1996, making it the second largest cigar manufacturer in the world. Of this figure, over 1.5 billion were exported to over 100 countries. The major markets for Dutch cigar export are within the European Union and they account for about 85% of the total exports. The leading export destinations in order of importance are: France, the United Kingdom, Germany and Belgium.

During 1996 Holland imported over 21 million cigars and cigarillos from the Canary Islands and nearly 7 million from the Dominican Republic. In total the country imported 50 million cigars and cigarillos.

AGIO

The Agio brand of small cigars is very popular in Europe and is made with tobaccos from the Cameroon, Java and Sumatra. Its popular Mehari family of cigars comes in three different types: Mild & Light (light Connecticut wrappers with a light touch of Burley), Brasil (dark Brazilian wrapper leaves with a full-flavoured filler mixture) and Cameroon (with Cameroon wrappers).

NAME	SIZE	RING GAUGE
Mehari's Sumatra	10.2cm/4in	23
Mehari's Mild	10.2cm/4in	23
Senoritas Red Label	10.2cm/4in	21
Biddies Brasil	8.3cm/3¼in	20
Filter Tipped	7.6cm/3in	21

HENRI WINTERMANS

In December 1996, the Scandinavian Tobacco Group of Companies (ST), Copenhagen, acquired Henri Wintermans Sigarenfabrieken B.V. from the British-American Tobacco Company. This merger created the largest cigar manufacturer in Europe, with a combined annual output in the region of one billion cigars. The new company, which retains the internationally-renowned name of Henri Wintermans, is one of the leading cigar exporters in the world, with a global distribution network spanning over 100 countries. The headquarters of Henri Wintermans is in Eersel, Holland, and the company employs almost 2,000 people in factories, warehouses and tobacco-processing plants, stretching from the Dominican Republic through Europe.

The new Henri Wintermans range, which includes Small Cigars, Half Coronas and Coronas de Luxe, represents the best Dutch tradition of blending and cigar-making. The company also manufactures Slim Panetelas, Slim Señoritas, Excellentes and Slim Coronas. The tobaccos are

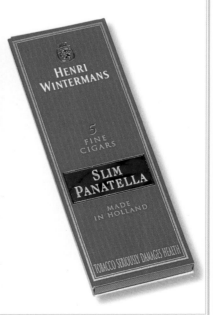

BELOW: Henri Wintermans is one of the leading cigar exporters in the world, with a highly regarded and well-known range of cigars.

NAME	SIZE	RING GAUGE
Excellente	16cm/6¼in	42
Long Panetela	13cm/5⅛in	30
Slim Panetela	13cm/5⅛in	30
Corona de Luxe	11.5cm/4½in	42
Half Corona	9.5cm/3¾in	30
Scooters	9cm/3½in	20

double-aged: the leaves are left for two years to mature and the rolled cigars are then aged for a second time, over cedarwood shelves, to create a round and full-bodied taste. The range is aimed at consumers in the medium to high end of the international cigar market, who may have graduated from mainstream brands, but who are not yet regular smokers of premium brands.

The company also produces the Café Crème range, which is the best-selling small cigar in the world. Variants include Café Crème Mild, Café Crème Rich Aroma, Café Crème Noir and Café Crème Filter Tip.

FROM LEFT TO RIGHT: Corona de Luxe, Slim Panetela Cigar, Small Cigar (part of the new Wintermans range), Half Corona.

SCHIMMELPENNINCK

In 1924, two brothers by the name of Van Schuppen became partners in their uncle's firm, Geurts. At that time Geurts was a small cigar factory employing 21 people. This company, Geurts & Van Schuppen, started making cigars under the Schimmelpenninck brand. Schimmelpenninck is the name of a famous Dutchman, who was governor of Holland in the early 19th century. The factory was just one of the many local family companies which existed at that time, manufacturing handmade cigars to its own specifications. Until 1930 it showed very little expansion.

In 1931, the Van Schuppen brothers started a policy of mechanization, and this, together with more attractive packaging and improved sales methods, resulted in a growth to one of the larger factories in Holland. After World War II the company linked with the firm of Carreras in 1963 and was then bought by Rothmans in 1972. Since that time exports of Schimmelpenninck cigars have increased strongly and the brand is sold in more than 130 countries around the world.

ABOVE: *Schimmelpenninck cigars are attractively packaged and sold worldwide.*

Name	Size	Ring gauge
Duet Brazil	14.3cm/5⅝in	27
Vada	9.8cm/3⅞in	30
Florina	9.8cm/3⅞in	26
Mono	8.6cm/3⅜in	27
Media	7.6cm/3in	26
Havana Mild	7.6cm/3in	26
Nostra	7.3cm/2⅞in	27

*FROM LEFT TO RIGHT: Half Corona, Havana Mild,
Havana Sigue, Media, Mini Cigar, Mini Cigar Mild, Swing,
Duet Panetela I, Grande.*

SWITZERLAND

VILLIGER

Villiger is one of the most popular machine-made small cigars in the world. The company was founded in Switzerland in 1888 by Jean Villiger and has grown to be a large international business. The Villiger family descendants still oversee the production of over 400 million cigars that are made in Switzerland, Germany and Ireland. The distinctive cigars are exported to over 70 countries throughout the world. Some of the cigars use homogenized tobacco, but many of them are manufactured with fine aged tobaccos imported from Cameroon, Java, Mexico, Columbia and the Dominican Republic.

FROM TOP TO BOTTOM: Villiger Export (round), Villiger Export (pressed).

NAME	SIZE
Curly	17.8cm/7in
Tipped Villiger-Kiel Mild	16.5cm/6½in
Villiger Export Kings	13cm/5⅛in
Tipped Bunte	11.4cm/4½in
Villiger Export	10.2cm/4in
Villiger Premium No. 4	10.2cm/4in
Braniff No. 1	8.9cm/3½in
Braniff Cortos Filter Light	7.6cm/3in

PUERTO RICO

Puerto Rico is a self-governing island, set in the West Indies to the east of the island of Hispaniola. It was discovered by Christopher Columbus in 1493. The island was colonized by the Spanish from 1510 until 1898, when it was ceded to the USA following the Spanish-American War. It became a US territory in 1917 and a Commonwealth in 1952. In recent years, many American companies have been attracted to Puerto Rico by various tax incentives. Among them is the Consolidated Cigar Corporation, which manufactures several brands of mass-market cigars on the island.

FROM LEFT TO RIGHT: Dutch Masters Cameroon Elite, Dutch Masters Cadet, El Producto Bouquet, El Producto Panetela.

THE UNITED STATES

The Swisher International Group was founded in 1861 and is the leading manufacturer and seller of cigars in the world. The company sells mass-market large cigars, premium cigars and little cigars. Mass-market large cigars use natural leaf wrappers or reconstituted leaf wrappers. Premium cigars are slightly more expensive, and are generally made with a natural wrapper, binder and long filler. General Cigar Company produces a range of mass-market cigars as well as their handmade premium cigars. Popular brands include Garcia Vega and White Owl.

FROM LEFT TO RIGHT: *King Edward Invincible de Luxe, King Edward Imperial (produced by Swisher International), Garcia Vega, White Owl (produced by the General Cigar Corp.).*

DIRECTORY OF CIGAR SHOPS

UNITED KINGDOM

LONDON

Alfred Dunhill of London
30 Duke Street
London SW1Y 6DL
Tel: 0171 499 9566
Fax: 0171 499 6471

Alfred Dunhill of London
5 Royal Exchange
Cornhill
London EC3V 1LL
Tel: 0171 623 9977
Fax: 0171 623 9445

Benson & Hedges
13 Old Bond Street
London W1X 4QP
Tel: 0171 493 1825
Fax: 0171 491 2276

Burlington Bertie
57 Houndsditch
London EC3A 8AA
Tel: 0171 929 2242
Fax: 0171 929 2232

Davidoff of London
35 St James's Street
London SW1A 1HD
Tel: 0171 930 3079
Fax: 0171 930 5887

G. Smith & Son
74 Charing Cross Road
London WC4H 0BG
Tel: 0171 836 7422
Fax: 0171 836 7422

Harrods Cigar Room
Knightsbridge
London SW1X 7XL
Tel: 0171 730 1234

Havana Club
165 Sloane Street
London SW1X 9QF
Tel: 0171 245 0890
Fax: 0171 245 0895

J. J. Fox of St James's
19 St. James's Street
London SW1A 1ES
Tel: 0171 930 3787
Fax: 0171 495 0097

Jayems
125 Victoria Street
London SW1E 5LA
Tel: 0171 828 1472

Sautter of Mayfair
106 Mount Street
London W1Y 5HE
Tel: 0171 499 4866
Fax: 0171 499 4866

The Segar & Snuff Parlour
27a The Market
Covent Garden
London WC2
Tel: 0171 836 8345

Selfridges Cigar Department
400 Oxford Street
London W1A 1AB
Tel: 0171 629 1234
Fax: 0171 491 1880

Shervingtons
337 High Holborn
London WC1V 7PX
Tel: 0171 405 2929
Fax: 0181 803 8887

Walter Thurgood
161–162 Salisbury House
London Wall
London EC2M 5QD
Tel: 0171 628 5437
Fax: 0171 930 5887

Wards of Gresham Street
60 Gresham Street
London EC2V 7BB
Tel: 0171 606 4318
Fax: 0171 606 4318

THE SOUTH

Burkitts
117 Church Road
Hove
East Sussex BN3 2AF
Tel: 01273 731351

Coster & Son
52 High Street
Marlow
Bucks SL7 1AW
Tel: 01628 482045
Fax: 01628 488998

Dome Tobacconist
2 Dome Building
The Quadrant
Richmond
Surrey TW9 1DT
Tel: 0181 940 3001

Harrison & Simmonds of
Bedford
80 High Street
Bedford MK40 1NN
Tel: 01234 266711
Fax: 01234 266711

M. Shave (Newbury)
1 The Arcade
Newbury
Berks RG14 5AD
Tel: 01635 46148

M. Shave of Reading
4 Harris Arcade
Reading
Berks RG1 1DN
Tel: 01734 595670

WALES AND THE WEST

C. A. Wrigley
35 Baldwin Street
Bristol BS1 1RG
Tel: 01179 273676

Frederick Tranter
5 Church Street
Abbey Green
Bath BA1 1NL
Tel: 01225 466197
Fax: 01225 466197

Lewis Darbey & Co
28–32 Wyndham Arcade
Mill Lane
Cardiff CF1 1FJ
Tel: 01222 233443

McGahey the Tobacconist
245 High Street
Exeter EX4 3NZ
Tel: 01392 496111
Fax: 01392 496113

MIDLANDS AND EAST ANGLIA
Churchills of Norwich
32 St Andrews Street
Norwich
Norfolk NR2 4AF
Tel: 01603 626437

Gauntleys of Nottingham
4 High Street
Nottingham NG1 2ET
Tel: 01159 417973
Fax: 01159 509519

Harrison & Simmonds of Cambridge
17 St John's Street
Cambridge CB2 1TW
Tel: 01223 324515
Fax: 01223 324515

J. M. Edwards
24 Fountain Street
Hanley
Stoke on Trent
Staffs SD1 1LD
Tel: 01782 281416
Fax: 01782 204246

John Hollingsworth & Son Ltd
5 Temple Row
Birmingham B2 5LG
Tel: 0121 236 7768
Fax: 0121 236 3696

John Hollingsworth & Son Ltd
97 High Street
Solihull B91 3SW
Tel: 0121 705 4549
Fax: 0121 705 4549

Lands (Tobacconists) Ltd
29 Central Chambers
Henley Street
Stratford upon Avon
Warwickshire CV37 6QN
Tel: 01789 292508

Tobacco World (Cheltenham)
Unit F7
Regent Arcade
Cheltenham
Gloucestershire GL50 1JZ
Tel: 01242 222037
Fax: 01242 222037

THE NORTH
Arthur Morris Ltd
71 Bradshawgate
Bolton BL1 1QD
Tel: 01204 521340
Fax: 01204 521340

Birchalls of Blackpool Ltd
10 Clifton Street
Blackpool FY1 1JP
Tel: 01253 24218
Fax: 01253 291659

C. Aston
23 Minden Parade
Bury BL9 0QD
Tel: 0161 764 2026

Greens Tobacconist
37 The Headrow
Leeds LS1 6PU
Tel: 0113 244 4895
Fax: 0113 245 9417

James Barber
33 Kirkgate
Otley LS21 3HN
Tel: 01943 462603
Fax: 01943 468770

Marhaba Newsagents
39 Cross Street
Manchester M2 4LE
Tel: 0161 834 9744

Tobacco World of Chester
78 Northgate Street
Chester CH1 2HT
Tel: 01244 348821
Fax: 01244 348821

SCOTLAND
Dallings of Ayr
5 Burns Statue Square
Ayr KY7 1SU
Tel: 01292 265799
Fax: 01292 265799

Herbert Love
31 Queensferry Street
Edinburgh EH2 4QU
Tel: 0131 225 8082
Fax: 0131 225 8082

House of Gowrie
90 South Street
Perth PH2 8PD
Tel: 01738 626919
Fax: 01738 626919

The Pipe Shop
92 Leith walk
Edinburgh EH6 5HB
Tel: 0131 553 3561
Fax: 0131 555 2591

Robert Graham & Co
71 St Vincent Street
Glasgow G2 5TF
Tel: 0141 221 6588
Fax: 0141 221 6588

Steve Silletts
166 King Street
Aberdeen AB24 5BD
Tel: 01224 644455

Tobacco House
9 St Vincent Place
Glasgow G1 2DW
Tel: 0141 226 4586
Fax: 0141 226 4586

UNITED STATES
NEW YORK
Alfred Dunhill Ltd
420 Park Avenue
New York, NY
Tel: 212 753 9292

Arnold's Cigar Store
323 Madison Avenue
New York, NY
Tel: 212 697 1477

Barclay-Rex Inc.
70 E. 42nd Street
New York, NY
Tel: 212 962 3355

The Big Cigar Company
193 A Grand Street
New York, NY
Tel: 212 966 9122

Cigar Emporium
541 Warren Street
New York, NY
Tel: 518 828 5014

Davidoff of Geneva
535 Madison Avenue
54th Street
New York, NY
Tel: 212 751 9060

De La Concha
Tobacconists
1390 Avenue of the
Americas
New York, NY
Tel: 212 757 3167

J.R. Tobacco
11 E.45th Street
New York, NY
Tel: 212 983 4160

Nat Sherman Inc.
500 Fifth Avenue
New York, NY
Tel: 212 246 5500

North Cigar Lounge
483 Columbus Avenue
New York, NY
Tel: 212 595 5033

The Smoking Shop
45 Christopher Street
New York, NY
Tel: 212 929 1151

OUTSIDE NEW YORK
Diebels Sportsmens
Gallery
426 Ward Parkway
Kansas City, KS
Tel: 800 305 2988

Georgetown Tobacco
3144 M North West
Washington, DC
Tel: 202 338 5100

Holt Cigar Co. Inc.
1522 Walnut Street
Philadelphia PA
Tel: 800 523 1641

The Humidor Inc.
6900 San Pedro Avenue
San Antonio TX
Tel: 210 824 1209

Jack Schwartz Importers
175 W. Jackson
Chicago, IL
Tel: 312 782 7898

J. R. Tobacco of America
Inc.
I-95 at Route 70
Selma AL
Tel: 800 572 4427

The Owl Shop
268 College Street
New Haven, CT
Tel: 203 624 3250

The Pipe Squire
346 Coddrington Center
Santa Rosa, CA
Tel: 707 573 8544

Rich Cigar Store Inc.
801 Southwest Alder St
Portland, OR
Tel: 800 669 1527

Tinder Box Santa Monica
2729 Wilshire Boulevard
Santa Monica, CA
Tel: 310 828 4511

CANADA
BRITISH COLUMBIA
Casa de Malahato
@ Malahat Chalet
265 Malahat Drive
Malahat V0R 2L0
Tel: (604) 478 0812

Sheffield & Sons
Tobacconist
320-A 4741 Lakelse Avenue
Terrace V8G 1R5
Tel: (604) 635-9661

VANCOUVER
Alpha Tobacco
927 Denman Street
Vancouver V6G 2L9
Tel: (604) 688 1555

R J Clarke Tobacconist
3 Alexander Street
Vancouver V6A 1B2
Tel: (604) 681 8021

Vancouver Cigar Company
1938 Broadway
Vancouver V6J 1Z2

TORONTO
Groucho & Company
150 Bloor St W
Toronto M5S 2XY
Tel: (416) 922 4817

Havana House
87 Avenue Road
Toronto M5R 3R9
Tel: (416) 927 9070

QUEBEC
Davidoff
1452 rue Sherbrooke W
Montreal H3G 1K4
Tel: (514) 289 9118

La Casa del Habon
1434 rue Sherbrooke W
Montreal H3G 1K4
Tel: (514) 849 0037

AUSTRALIA

ADELAIDE

Tunney's
38–40 Grote Street
Tel: (08) 8231 5720

MELBOURNE

Alexander's Cigar Divan
at Crown Towers
8 Whiteman Street
Southbank
Tel: (03) 9292 7842

Baranow's Fine Cigars
P. O. Box 29
Preston Vic 3072
Tel: (03) 9479 6579
E-mail: Cigar1 @
MSN.com
Web site: HYPERLINK
HTTP://WWW.aml.com.a
ut/Cigar1.htm
HTTP://WWW.aml.com.a
ut/Cigar1.htm

Benjamin's Fine Tobacco
Shop 10, Strand Central
250 Elizabeth Street
Tel: (03) 9669 2879
E-mail: Bentob @
Netspace.net.au

J & D of Alexander's
7A–459 Toorak Rd
Toorak 342
Tel: (61) 9827 1477

SYDNEY

Alexander's Cigar Divan
at Pierpont's
Hotel Intercontinental
117 Macquarie Street
Tel: (02) 9252 0280

Sol Levy
713 George Street
Tel: (02) 9211 5628

NEW ZEALAND

Havana House, Cigars
Limited
11–19 Customs ST. W.
Auckland
Tel: 64 9 357 0037

Imperial Tobacco
295-A The Terrace
Wellington
Tel: 64 4 801 9002

SOUTH AFRICA

Wesley's
Golden Acre Plaza
Level 7, Cape Town
Tel: 27 21 21 5090

Wesley's
The Rosebank Mall
170
Johannesburg
Tel: 27 11 6333 2510

PICTURE CREDITS

T = top, b = bottom; l = left, r = right, c = centre

P1 Dave Jordan; p2 John Freeman; p3 Don Last; p6 b Jon Wyand; p10 bl ET Archive; p10 r: Chris Sharp/South American Pictures; p11 cl Visual Arts Library; p12 tr Peter Newarkís Pictures; p12 bl AKG; p13 bl Topham Picture Point; p13 br Mary Evans Picture Library; p 13 tr John Wyand; p14 bl Images Colour Library; p15 tl Images Colour Library; p15 tr John Wyand; p18 bl Bridgeman Art Library; p18 tr AKG; p18 cr ET Archive; p19 t, p19 bl and p19 br Peter Newarkís Pictures; p20 South American Pictures; p21 tr Mary Evans Picture Library; p21 tl Images Colour Library; p21 b Elliot Erwitt/Magnum; p 21 tl James Nachtwey/Magnum; p21 tr David Alan Harvey/Magnum; pp22/23 Images Colour Library; p26 tr John Wyand; p27 bl David Alan Harvey/Magnum; p27 tr Images Colour Library; p28 tl Mary Evans Picture Library; p28 br South American Pictures; p28 bl South American Pictures; p29 bl Susan Meiselas/Magnum; p29 tr Edward Parker; p30 bl John Wyand; p32 tl Murray Rothmans; p32 tr John Wyand; p 32 bl John Wyand; p33 bl John Wyand; p33 br Tania Jovanovic; p34 t, c and br Tania Jovanovic; p35 tr John Wyand; p35 bl John Wyand; p35 br Tania Jovanovic; p37 Tomas Sennet/Magnum; p42 bl John Freeman; p42 br Don Last; p43 br John Wyand; pp44/45 Dave Jordan; p46 tc and bl Alfred Dunhill; p48 Don Last; p49 t and

b Don Last; pp52/3 Dave Jordan; p61 tc Peter Newark's Historical Pictures; p62 tl Tania Jovanovic; p62 c Mary Evans Picture Library; p64 t Don Last; p64 b Dave Jordan; p69 t Dave Jordan; p72 br Dave Jordan; p72 c Mary Evans Picture Library; p73 br Mary Evans Picture Library; p73 tr Burt Glinn/Magnum; p73 cl Kobal Collection; p74 cr Mary Evans Picture Library; p74 tl Hulton Picture Library; p74 tr Peter Newark's Pictures; p75 Bridgeman Art Library; p78 t Images Colour Library; p78 b Images Colour Library; p82 t Dennis Stock/Magnum; p82b Images Colour Library; p83 b Murray Rothman's; p84 bl and br Tania Jovanovic; p85 tl and br John Wyand; p97 Tania Jovanovic; p102 Visual Arts Library; p114 tr Images Colour Library; p114 bl Murray Rothmans; p115 br Murray Rothman's p116 t Murray Rothman's; p130 Murray Rothmans; p157 Murray Rothmans; p183 b Murray Rothman's; p188 Robert Francis/South American Pictures; p217 Colour Images Library; p220 Colour Images Library; p223 t James Nachtewey/Magnum; p223 b Susan Meiselas/Magnum; p228 Colour Images Library; p230 Colour Images Library; p231 Alex Webb/Magnum; p239 Fine Art Photographic Library. The Publishers would also like to thank John Hall for the cigar label on p56 b and Amoret Tanner, for lending some of her memorabilia for photography, including the ephemera on p62.

INDEX